MA
SOUTH EAST
ASIA

Charles M. Francis

POCKET PHOTO GUIDE

B L O O M S B U R Y
LONDON · OXFORD · NEW YORK · NEW DELHI · SYDNEY

Bloomsbury Natural History
An imprint of Bloomsbury Publishing Plc

50 Bedford Square
London
WC1B 3DP
UK

1385 Broadway
New York
NY 10018
USA

www.bloomsbury.com

BLOOMSBURY and the Diana logo are trademarks of
Bloomsbury Publishing Plc

First published by New Holland UK Ltd, 2001 as *A Photographic Guide
to Mammals of South-East Asia*
This edition first published by Bloomsbury, 2016

British Library Cataloguing-in-Publication Data
A catalogue record for this book is available from the British Library.

Library of Congress Cataloguing-in-Publication data has been applied for.

ISBN: PB: 978-1-4729-3797-1
ePDF: 978-1-4729-3799-5
ePub: 978-1-4729-3796-4

2 4 6 8 10 9 7 5 3 1

Designed and typeset in UK by Susan McIntyre
Printed in China

FSC
www.fsc.org

MIX
Paper from
responsible sources
FSC® C104723

To find out more about our authors and books visit www.bloomsbury.com.
Here you will find extracts, author interviews, details of forthcoming events
and the option to sign up for our newsletters.

CONTENTS

INTRODUCTION

Mammals are familiar to everybody, as they include domestic animals such as dogs, cats, horses and cattle as well as, of course, ourselves – humans. However, fewer people know about the tremendous diversity of wild mammals. Over 4,400 species occur around the world. South-east Asia is particularly rich in mammals with over 500 species in the region covered by this book, including wild cattle, elephants, rhinoceros, wild pigs, wild cats, bears, monkeys and many smaller species such as rodents, shrews and bats.

Watching mammals presents more challenges than watching birds, since many species are small and nocturnal, and difficult to observe and identify. In South-east Asia, even larger mammals can be difficult to find, as their natural habitats are thick tropical forests. Nevertheless, a keen observer can locate many mammals. The loud morning songs of gibbons ring across the forests in many areas. Many monkeys and squirrels can be seen during a walk through forests or gardens. Civets, flying squirrels, lorises and other mammals may be found during a night walk through lowland rainforest with a spotlight. Larger mammals such as cattle, tapirs, or bears can be located by tracks or signs, or seen from hides near salt licks. Bats can often be seen at roost in caves, or at dusk as they hunt for insects.

This book provides an introduction to the wild mammals of South-east Asia, including the countries of Myanmar (Burma), Thailand, Laos, Vietnam, Cambodia, Malaysia, Singapore, Brunei, and western Indonesia including Sumatra, Java, Bali, and Borneo. Most species of larger mammals are included, which should enable identification of many animals encountered in these groups. For smaller animals, such as rodents, insectivores and bats, only representative species are included, because there is not space in a book of this size to describe all species.

Much remains to be learned about mammals in South-east Asia. Since 1990, several new species of large mammals have been discovered that were previously unknown to science, including Saola, Large-antlered Muntjac, Puhoat Muntjac and Annamite Muntjac from the Annamite mountains in Laos and Vietnam, and Leaf Muntjac from Northern Myanmar. Many more species of small mammals have also been discovered, including a new striped rabbit, several new rodents and many new bats. Many areas have never been thoroughly surveyed for small mammals, and no doubt further species await discovery.

Unfortunately, this tremendous diversity of mammals is also at risk. Several larger mammals are threatened with imminent extinction. Schomburgk's Deer, which formerly occurred in lowland swamps in Thailand, is believed to be already extinct. The continued existence of Kouprey is in doubt, and the newly discovered Saola is under heavy hunting pressure in Laos and Vietnam. The Asian One-horned Rhinoceros is now considered extinct in Vietnam, while a tiny population of about 50 survives in western Java. The Asian Two-horned Rhinoceros has disappeared from much of its range. Many primates, including Orangutans and Proboscis Monkeys, have experienced substantial declines. Tiger populations in many regions are probably too small to be sustainable.

The greatest single threat to most Asian mammals is loss of habitat. Until recently, most of South-east Asia was covered in continuous forest. The island of Borneo formed one of the largest continuous blocks of tropical rainforest in the world. Now, much of this has been cleared or severely degraded. Less than 10% of the land area in most jurisdictions is set aside as parks or reserves, and much forest that remains has been heavily logged. Some large mammals can adapt to logged forest, especially those that traditionally fed in forest openings or grassy areas along river banks. However, many smaller mammals with more specialized habitat requirements may be unable to adapt. Furthermore, logged forests are vulnerable to encroachment and fires. In the past two decades, numerous large forest fires have burned millions of hectares of logged forest in Borneo and Sumatra. These burned-over forests, even if protected from further disturbance, will take decades or even centuries to recover. Some areas where seed sources are gone may never regain their original diversity.

Another threat to many mammals is hunting. Although some mammals have been traditionally hunted for centuries, growing human populations and the widespread availability of guns have greatly increased hunting pressure. At the same time, with loss of habitat, many mammal populations have become smaller and more vulnerable to hunting. Development of roads and agricultural areas around forest reserves has increased opportunities for poachers, and the demand for parts of some species, such as rhinoceros horns, pangolin scales, or tiger bones, for so-called 'medicinal' purposes, has greatly increased the profits associated with illegal hunting. Unless drastic efforts are made to curtail the illegal wildlife trade, many larger mammals will become extinct in the near future.

The loss of a species represents an irreversible loss of millions of years of evolution. Humans have a moral obligation to prevent the extinction of any species, both for the sake of the animals themselves, and so that our children and their children can see and appreciate them. Also, wild mammals represent an inestimable genetic resource. The wild ancestors of many economically important domestic mammals are still found in Asia today – but they are also threatened. Wild populations could prove invaluable for improving domestic stocks through cross-breeding programmes, or for developing new domestic breeds – if they are preserved. Finally, many wild mammals are of great ecological importance. For example, several species of fruit bats are key pollinators of such economically important trees as durians, kapok and mangroves. Other bats, as well as squirrels and many larger mammals including monkeys, bears and even elephants, are important seed dispersers and play a key role in forest regeneration. Insectivorous bats eat hundreds of millions of insects every night. Wild carnivores help control rodent populations, thus maintaining ecological balance.

The continued existence of this tremendous diversity of native mammal species in South-east Asia, and the natural habitats where they live, is dependent on the stewardship and care of the people living in the region. One of the primary goals of this book is to increase awareness of the fascinating diversity of wildlife in the region, in the hope that improved understanding will lead to better care and conservation of wild animals into the future.

HOW TO USE THIS BOOK

This guide has been compiled to help with identification and increase understanding of mammals in South-east Asia. It should allow identification of many of the larger mammals to species, and most smaller mammals to family or genus.

Small thumbnail sketches are included to help determine and locate the group to which a mammal might belong. Brief descriptions are provided for each order. Photographs and text descriptions should be examined and compared to check identification. The approximate size of each mammal is indicated by one of three measurements. For most mammals, the total length (TL) from the tip of the nose to the tip of the tail (excluding long hairs) is given. Note that with their back hunched or tail curled, many animals may look shorter. For most larger mammals, shoulder height (SH) is given, while for bats, the forearm measurement (FA) is provided.

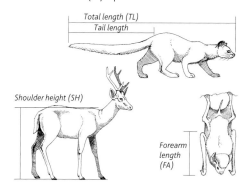

WHAT IS A MAMMAL?

Mammals are warm-blooded vertebrates distinguished by having fur or hair, and by giving birth to live young (usually), which they suckle on milk. Some mammals may be confused with other animals. Whales and dolphins have virtually no hair, and swim like fish, but are warm-blooded, give birth to live young which they suckle on milk, and breathe air. They are thought to have descended from ancestors that fed on land and walked on four legs. Bats are sometimes confused with birds because they can fly, but they have fur, not feathers, teeth instead of a beak, and in all other ways are clearly mammals. Pangolins superficially resemble reptiles because of their scales, but these are formed from densely packed hairs, and pangolins share all the other features of mammals.

CLASSIFICATION AND NAMES

Mammal species are classified into genera (plural of genus), families and orders. All mammals have a scientific name, used by scientists throughout the world, indicating the genus and species. These names follow strict rules, but may change based on new information. English names are less well standardized and vary among sources. Scientific and English names used in this book have been largely updated to follow Francis (2008) *A Field Guide to the Mammals of South-East Asia*, published by Bloomsbury.

KEY TO COLOURED TABS

 Pangolins

 Moonrats

 Shrews

 Treeshrews

 Colugos

 Fruit Bats

 Insectivorous bats

 Loris & Tarsier

 Monkeys

 Gibbons & Orangutan

 Wild dogs

 Bears

 Otters, martens & badgers

 Civets

 Mongooses

 Cats

 Dolphin & Dugong

 Elephants

 Tapir & rhinoceros

 Wild pigs

 Deer

 Cattle

 Saola

 Squirrels

 Flying squirrels

 Mice & rats

 Porcupines

 Rabbits

GLOSSARY

Aquatic Living in or near water.
Arboreal Adapted for life in trees.
Carnivorous Eating mainly meat, and preying on other animals for food.
Diurnal Active mainly during daylight hours.
Endemic Found only in a particular region, and nowhere else in the world.
Frugivorous Feeding mainly on fruit.
Insectivorous Feeding mainly on insects.
Nocturnal Active mainly at night.
Omnivorous Feeding on both animals and plant material.
Pedicel The bony base of a deer's antler
Terrestrial Active mainly on the ground.
Tine A branch on the antler of a deer.

MAMMALS IN SOUTH-EAST ASIA

The distribution of mammals in South-east Asia is influenced by climate, vegetation, altitude and history. The region covered by this book can be divided into three major subregions. In the north, the Himalayan subregion, with cooler climates and higher altitudes includes the northern tip of Myanmar. Several mammal species of northern Asia and Europe inhabit this area, but are not emphasised in this book, as they are more typical of other regions.

The Indochinese subregion includes the remainder of Myanmar, as well as Laos, Vietnam, Cambodia and Thailand except the extreme south. It has monsoon climates, with a distinct dry and wet season. Many forests, including dry dipterocarp forest, are semi-deciduous with some trees losing their leaves in the dry season. Wet evergreen forest occurs in the hills and some coastal areas.

The Sunda subregion includes peninsular Thailand (south of the Isthmus of Kra), Malaysia, Sumatra, Java and Borneo. The Sunda region is generally much wetter, with a monthly average of at least 100 mm of rain throughout the year, leading to growth of rich lowland rainforests. Although this subregion is divided into islands, the seas between them are relatively shallow and were exposed as dry land whenever sea levels dropped during ice ages, most recently about 10,000 years ago. As a result, the islands share many plants and animals. The division from the Indochinese region is very marked botanically, with 375 Sundaic genera of plants reaching their northern limit, and 200 genera of Indochinese plants reaching their southern limit there. Several mammal species show similar limits in this region.

Tall lowland rainforests, that once covered most of the Sunda subregion and part of the Indochinese region, are the richest habitats for mammals. Hill and montane forest supports fewer species, although some mammals are more common in hill forests, such as Siamang, Serow, Saola and several squirrels and rats. Monsoon forests support a distinctive fauna, although they are generally less rich than rainforests. Limestone outcrops occur throughout the region, and often have extensive cave systems that shelter large colonies of bats. A few rodents and other mammals, including Francois's Langur, are restricted to limestone areas. Coastal mangrove forests are important for some species. Only a few mammals have adapted to the open scrub and field habitats that are left following destruction of the original forests.

HOW TO FIND MAMMALS

Seeing many diurnal mammals, such as squirrels, monkeys and apes, requires only a visit to the appropriate habitat, a good pair of binoculars, and some patience. The most suitable binoculars are 7–8 x magnification with an ocular (the lens at the end) at least 35 to 40 mm in diameter (e.g. 7 x 35 or 8 x 40). Smaller binoculars, although easy to carry, are not bright enough to see details inside the forest.

Monkeys, gibbons and some tree squirrels can be found by their loud calls or by the crashing of branches as they move from tree to tree. Ground squirrels and treeshrews can be found by listening for rustling leaves or looking for movement near to the ground. Larger mammals can be found by their footprints and other signs. A boat trip along a river in the late afternoon or early morning can be a good way to see monkeys such as Proboscis Monkeys and macaques, and possibly deer and other large mammals coming to drink.

Nocturnal mammals can be found by walking quietly along trails or roadsides at night, with a headlamp or spotlight to look for eyeshine. Many animals are less wary of people at night, and it is often possible to get much closer than during the day. Species such as lorises, flying squirrels and some civets can be seen in the treetops, while deer, civets and cats may be seen on the ground. Larger mammals such as pigs, deer or elephants can sometimes be found by driving along secondary roads through forest, with a spotlight. Some reserves have hides where visitors can watch over a water hole or salt lick for mammals.

Bats can often be observed at roost, or as they fly out from their roosts in houses, trees or caves, but only a few species can be identified in flight. Bat researchers use various methods to capture bats for identification, including mist nets and harp traps. A 'bat detector' transforms the echolocation calls of insectivorous bats to frequencies that humans can hear, or displays their calls on computer screens. This enables researchers to study and identify bats without disturbing them, in the same way birders can identify birds by their songs. Most shrews, mice and rats must be captured for identification, using a variety of small mammal traps. To capture a full range of species, traps must be set from ground level to the treetops, using different baits. Experience is needed to use traps successfully without injuring the animals, and in most areas, special permits are needed to capture mammals.

Do not be discouraged if you cannot always identify what you see. The appearance of an animal may change depending on the angle and the lighting, and many species vary in colour, both within a population and among areas. Even experts cannot identify some mammals in the field without capturing and studying them. It is still possible to enjoy the experience of seeing a wild mammal even if it cannot be precisely identified. This experience can be greatly enhanced by observing its behaviour, watching it interact with other animals and its environment.

WHERE TO FIND MAMMALS

Numerous parks, nature reserves and other protected areas have been designated in each country in the region, many of which still support a good variety of mammals. In this book, it is only possible to highlight a few of the more important or more accessible reserves or parks in the region. This selection is inevitably rather arbitrary, and many other sites may also be good for mammals.

Myanmar currently has 39 designated or proposed protected areas, but access to many is limited, especially to foreign tourists. The map highlights four of the most important for mammals. After years of isolation, all are becoming more accessible to foreigners.

Thailand has 60 national parks and 31 designated wildlife sanctuaries. Khao Yai is one of the most accessible sites, still supporting many primates, elephants and other types of mammals. Huai Kha Kheng and Kaeng Krachan are among the largest and richest areas for mammals.

Laos has 29 proposed or designated National Biodiversity Conservation Areas (NBCA), covering most habitats in the country, although visitor access is limited. One of the most important is

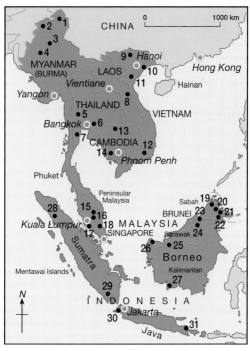

1 Hkakaborazi National Park
2 Tamanthi Wildlife Sanctuary
3 Chatthin Wildlife Sanctuary
4 Alaungdaw Kathapa National Park
5 Huai Kha Khaeng Wildlife Sanctuary
6 Khao Yai National Park
7 Kaeng Krachan National Park
8 Nakai Nam-Theun
9 Na Hang Nature Reserve
10 Cat Ba National Park
11 Cuc Phuong National Park
12 Cat Tien National Park
13 Kulem Promtep
14 Bokor
15 Taman Negara
16 Krau Wildlife Reserve
17 Endau Rompin
18 Kuala Selangor
19 Kinabalu Park
20 Sepilok
21 Kinabatangan River
22 Danum Valley
23 Batu Apoi Forest Reserve
24 Mulu National Park
25 Lanjak-Entimau Wildlife Sanctuary
26 Samunsam Wildlife Sanctuary
27 Tanjung Puting National Park
28 Gunung Leuser National Park
29 Bukit Barisan Selatan National Park
30 Ujung Kulon National Park
31 Baluran National Park

Nakai-Nam Theun NBCA, with Annamite endemics such as Saola, Large-antlered Muntjac and Red-shanked Douc.

Cambodia has a variety of parks and wildlife reserves, but many have not been surveyed recently, and the status of their wildlife is unknown. Bokor is relatively accessible and still supports many mammals, though some larger species are now rare. Kulen Promtep in the north may still support Kouprey (if they are not yet extinct) and Eld's Deer, as well as most other large mammals.

Vietnam has a range of reserves and protected areas. Cat Tien and Cat Loc have a wide range of mammals, including Black-shanked Douc, Gaur, Buff-cheeked Gibbon and rhinos. Cuc Phuong has Delacour's Langur, Owston's Palm Civet and many bats. Cat Ba is the only site for the golden-headed Cat Ba Langur. Na Hang is the last remaining site for Tonkin Snub-nosed Monkey.

The largest and most accessible park in **Peninsular Malaysia** is Taman Negara with a full complement of lowland forest species including tapirs, elephants and Gaur. Other large areas of lowland forest include Krau and Endau-Rompin. Coastal species such as Sundaic Silvered Langur and Smooth Otter can be seen at Kuala Selangor.

In **Sabah**, East Malaysia, Mount Kinabalu has all of the hill and montane species in Borneo, as well as some lowland species. Danum Valley has excellent facilities for visitors and researchers, with mature lowland forest and a rich variety of mammals. Sepilok is readily accessible from Sandakan, with semi-wild Orangutans and other mammals. Kinabatangan Wildlife Sanctuary supports Proboscis Monkeys, elephants, otters and wild cats, and nearby Gomantong Caves has spectacular colonies of bats. In **Sarawak**, Lanjak-Entimau is one of the largest and most important wildlife areas in Malaysia. Niah and Mulu both have spectacular limestone caves with numerous bats. Samunsam Wildlife Sanctuary is a good site for Proboscis Monkeys.

Brunei has extensive remaining forest. The most accessible is in Batu Apoi, where there is a fully-equipped research station.

In **Kalimantan**, Indonesian Borneo, there are over 70 proposed or designated protected areas, but many are not readily accessible or have been extensively damaged by illegal logging or fires. One of the best remaining is Tanjung Puting with coastal swamp and lowland forest and good populations of Orangutans and Proboscis Monkeys.

The richest area in **Java** is Ujung Kulon, with lowland and swamp forest and the last remaining population of Lesser One-horned Rhinoceros, as well as Leopards, gibbons and many other species. Baluran has dry forest and grasslands with Banteng and Javan Rusa.

In **Sumatra**, Gunung Leuser is one of the largest and richest parks, with most mammals including montane species. Bukit Barisan Selatan has a variety of habitats with large mammals including Tigers, elephants and tapirs.

Acknowledgements

I would like to thank the numerous individuals and agencies who supplied photographs for use or consideration in this book, including: Gerald Cubitt, Alain Compost, K. Fletcher, Louise Emmons, Indraneil Das, Bill Robichaud, Tony Lynam, Bruce Kekule, Thomas Geissmann, WWF Malaysia (especially Peter Teh), Manuel Ruedi, Peter Vogel, Anthony Lamb, Fauna & Flora International, Nature Focus, the Bruce Coleman Collection, the Harrison Institute, and the Mammal Society Slide Library. I also thank Ken Scriven, who was the driving force behind starting this book, Mike Unwin, who edited the book and patiently helped with locating all of the necessary photographs and Will Duckworth who kindly reviewed the draft text. My field research on mammals and birds in South-east Asia has been supported by the Wildlife Conservation Society, Asia programme, for which I would particularly like to thank Alan Rabinowitz and Josh Ginsberg as well as by Bird Studies Canada.

PANGOLINS Order Pholidota

With scales all over their body, pangolins superficially resemble reptiles, but are true mammals, with hairs between the scales (which are themselves formed from the same proteins as hair), and other features of mammals. Pangolins are found only in Africa and Asia. Two species occur in the region.

SUNDA PANGOLIN *Manis javanica* TL 75–102cm

Alain Compost

WWF Malaysia/Ken Scriven

Pangolins are easily recognized by their scaly body, long tail and long pointed muzzle. The tail is prehensile, and can be wrapped around trees or branches for support when climbing. The jaws lack teeth, but the long sticky tongue can extend up to 25cm to collect ants and termites. Up to 200,000 ants may be eaten in one meal. Long, powerful claws enable this species to dig into ant hills and tear open termite mounds. Tiny ears, thick skin and scales offer protection against the bites and stings of ants. The sharp-edged scales protect the pangolin against predators – it can roll into a ball, with its tail covering the belly and head, so that only the scales are exposed. Gives birth to one, occasionally two young at a time, which cling to base of mother's tail during travel. Found in eastern Myanmar, Thailand, central and southern Laos and Vietnam, Cambodia, peninsular Malaysia, Sumatra, Java and Borneo. Chinese Pangolin *M. pentadactyla* found in northern Myanmar, Thailand, Laos, Vietnam and much of China is similar, but with relatively longer front claws, larger ears, and fewer rows of scales on the tail (14 to 17 instead of about 30).

INSECTIVORES Order Insectivora

The Insectivores are thought to resemble the ancestral mammal from which all other mammals descended, with many relatively unspecialised teeth, and feeding largely on insects and other invertebrates. However, recent research indicates they are not all closely related to each other, and should be split into at least two orders. There are over 40 species currently recognized in the region of this book, divided into three families. The family Erinaceidae includes the moonrat and gymnures. The family Talpidae includes the moles, which spend much of their time underground and are rarely seen, though they may be located by the raised mounds along their burrows. Unfortunately no photographs are available. The remaining species are shrews in the family Soricidae, of which examples of the two most widespread genera are illustrated here.

MOONRAT *Echinosorex gymnurus* TL 45–70cm

WWF Malaysia/R Rajanathan

Moonrats are the largest insectivores in the region, weighing 0.5–2.0kg. In Borneo, most individuals are largely white with scattered black hairs, but elsewhere they tend to be black with white guard hairs on the back half of the body, and white on the front half. They inhabit lowland forests including some plantations. They forage on the ground, mainly at night, but sometimes during the day, eating worms, insects, millipedes, crabs, snails, amphibians and some plant matter. During day, they rest in hollow logs, under tree roots or in holes. Produce a strong odour, like rotten onions, from scent glands near base of tail. May produce two litters of usually two young per year. Occur in peninsular Thailand, Malaysia, Sumatra and Borneo.

SHORT-TAILED GYMNURE *Hylomys suillus* TL 14–16 cm

Short-tailed Gymnures are larger than most shrews, weighing 50–70g. They have a sharply pointed face and a distinctive short naked tail. Body fur is rusty brown above, paler below. Mainly terrestrial, hunting for food by both day and night. Probes through leaf litter on the forest floor, using its long snout to toss aside leaves. Diet includes worms, insects and other arthropods, and some fruit or other plant matter. Usually solitary, but sometimes in groups of two or three. May breed throughout the year, giving birth to two or three young per litter. Usually moves with short hops or leaps, but can travel quickly when disturbed. Found throughout the region, mainly in hill and montane forest, up to 3000m in Borneo and Sumatra, but down to nearly sea level in some areas including Tioman Island.

WHITE-TOOTHED SHREWS *Crocidura* spp. TL 8–20cm

Shrews are small active mammals with pointed muzzles and are fierce predators. They may eat other small mammals, frogs or lizards, but most species feed mainly on insects, worms and other invertebrates. White-toothed shrews of the genus *Crocidura* are found through much of Africa, Europe and Asia, and include over 170 species, more than any other genus of mammals. Distinguished from the similar *Suncus* shrews, which also have white teeth, by having eight instead of nine teeth on each jaw. At least 14 species occur in the region, separated by body size, the relative length of the tail (from 40 to 110% of head and body length), the amount of long bristles at the base of the tail, and the size and shape of the teeth. Malayan Shrew *Crocidura malayana*, found in peninsular Malaysia, is dark grey to blackish with a silvery gloss and a relatively thin tail about two-thirds its body length. It weighs 10–18g. Javan White-toothed Shrew *Crocidura orientalis*, known only from mountains around Cibodas in western Java, is similar but with a relatively longer tail lacking bristles at its base.

Manuel Ruedi

SAVI'S PYGMY SHREW *Suncus etruscus* TL 6–7cm

Peter Vogel

Savi's Pygmy Shrews are among the world's smallest mammals, weighing only 1.8–2.4g. Distinguished from young of other shrews by proportions, especially relatively small hind feet. Fur colour varies from a light greyish brown to dark brown. A fierce predator, this species eats two to three times its body weight in food per day and will readily tackle insects as large as itself. In Thailand, it apparently feeds mainly on termites that it captures on the forest floor. In Europe, gives birth to from two to six young after a gestation period of about 27 days. Found throughout southern Europe, the Mediterranean coast of Africa, central and southern Asia, mainland South-east Asia and Borneo, although there is some doubt whether these populations are really all the same species. The much larger House Shrew *Suncus murinus*, is pale grey all over with a thick tail. Its total length can be in excess of 20cm, with body mass from 40–110g depending on the population. Distinguished from similar-sized rodents by its pointed muzzle, very different teeth and thick tail that tapers to the tip. Young animals sometimes travel by caravanning, with the first holding its mother's tail, and the others each holding the tail of the one in front. Occurs throughout most of Africa and Asia. It is found mainly around human settlements and may have been accidentally introduced in many areas.

TREESHREWS Order Scandentia

Treeshrews have been variously considered primitive primates or insectivores, but are now usually placed in their own order. They superficially resemble squirrels, but have more pointed muzzles with many smaller teeth, not the chisel-like incisors of rodents. Five genera are recognized, of which one is confined to India, another to the Philippines, and three occur in the region covered by this guide. The genus *Tupaia* has 10 species, three of which are illustrated here. The genus *Dendrogale*, with one species in southern Indochina and another in the mountains of Borneo, is distinguished by a relatively long, thin tail with very short hairs. Pen-tailed Treeshrew *Ptilocercus lowii*, of southern Thailand, Malaysia, Sumatra and Borneo, is the only species that is mainly nocturnal. It is pale grey above, white below, and has a long naked tail except for a broad feather-like fringe of hairs at the tip.

COMMON TREESHREW *Tupaia glis* TL 30–45cm

WWF Malaysia

This treeshrew spends much time on the ground, on fallen logs, or in low bushes, but will readily climb trees. The upperparts and tail are grizzled reddish-brown, with a distinct pale mark on the shoulder; underparts paler. It is usually solitary, feeding on a variety of foods including insects and other arthropods, fruits, seeds and buds. In peninsular Malaysia, may breed at any time of year. It builds a nest in holes in fallen timber and similar sites, and gives birth to between one and three young after a 50 day gestation period. Young are able to leave the nest at 33 days, but may remain with the mother for some time afterwards. It is found from peninsular Thailand southwards, including Sumatra, Java and Borneo, but replaced in the rest of South-east Asia by Northern Treeshrew *Tupaia belangeri*, which is similar in appearance but less reddish. Lesser Treeshrew *Tupaia minor* (see below) is the only other *Tupaia* treeshrew on the mainland, but seven other species occur in Borneo, three in Sumatra and one in Java.

LESSER TREESHREW *Tupaia minor* TL 25–30cm

Rod Willams/Bruce Coleman Collection

Distinguished from most other treeshrews by relatively small body and long slender tail. Upperparts speckled olive-brown with a narrow buff shoulder stripe; underparts buffy, often with a reddish tinge. Mainly arboreal, climbing rapidly with ease along lianas and thin branches of trees. Most often seen in the lower middle storey a few metres off the ground, but may range up to 20m or more into the canopy. Diet includes a variety of fruit and insects. Gives birth to two (usually) or one young at any time of year, but with a peak of breeding, at least on the mainland, from May to July. Found in peninsular Thailand and Malaysia, Sumatra and Borneo. Slender Treeshrew *Tupaia gracilis*, confined to Borneo, is similar but lacks the reddish tinge and is slightly larger with a longer hind foot. Other treeshrews are larger and more terrestrial.

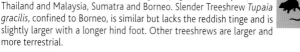

MOUNTAIN TREESHREW *Tupaia montana* TL 28–42cm

Gerald Cubitt

Similar to Common Treeshrew, but darker with reduced shoulder stripe, and smaller hind feet. Fur dark brown with fine reddish speckles. Feeds mainly on the ground, on a variety of plant and animal material. Found only in the mountains of north-east Borneo, usually above 900m. Other Bornean treeshrews differ in size or colour pattern, with several species showing patterns of reddish orange or black.

COLUGOS Order Dermoptera

Colugos, also called flying lemurs, although they are not closely related to true lemurs, are distinctive leaf-eating mammals with membranes connecting legs and tail, enabling them to glide from tree to tree. These gliding membranes have evolved independently from those of other gliding mammals such as flying squirrels or marsupial gliding possums, or from the wings of bats. Two species are recognized, one confined to the Philippines, and one elsewhere.

SUNDA COLUGO *Galeopterus variegatus* TL 50–70cm

WWF Malaysia/Dennis Yong

WWF Malaysia/Chew Yen Fook

Colugos occur in two colour phases. Females usually grey with reticulated pattern of white and black, while males may be grey, or reddish-orange with white patterns. The gliding membranes are the most extensive of any gliding mammal, connecting from the neck to the toes on both fore and hind legs and to the tip of the tail. When gliding between trees, can be distinguished from flying squirrels by having the tail connected as part of membranes, rather than hanging free. Recorded gliding up to 136m between trees, losing only 12m in height in the process. A mainly nocturnal species, spending the day usually at rest in a tree hole. Live in the tree-tops, being nearly helpless on the ground. This species lacks upper incisors, but the lower incisors are comb-shaped, with numerous elongated cusps on each tooth, that may be used to scrape leaves, or for grooming fur. The main diet is leaves and young shoots, although sometimes eat fruit. Give birth to a single young after gestation of about eight weeks. Reported from southern Myanmar and Thailand, south Cambodia and Vietnam, peninsular Malaysia, Sumatra, Java and Borneo.

BATS Order Chiroptera

Bats, the only mammals that can truly fly, are readily distinguished by their 'hand-wings', formed from elongated fingers with stretched skin between them. They are divided into two suborders. The Megachiroptera or fruit bats eat mainly fruit and nectar, and include a single family, the Pteropodidae, with about 25 species in the region. They have large eyes, simple ears, dog-like faces and short or lacking tails. Many species have a claw on their second finger, as well as their thumb, to aid climbing through trees. The Microchiroptera, which mainly eat insects, include more than 175 species in nine families in the region. Although all species can see well during the day, their eyes are generally small and they rely on echolocation (sonar) to find their way in the dark. Many parts of South-east Asia have been poorly surveyed for bats, and new species are regularly discovered.

FLYING FOXES Pteropus spp. FA 120–200mm

Flying foxes are the world's largest bats. *P. vampyrus* has a wingspan up to 1.5m and weighs up to 1.1kg. Resembles a small eagle in flight, but is distinguished by its slow flapping wingbeat and distinctive shape. Roosts in treetops in noisy, conspicuous colonies in rainforest, mangroves or even city parks. May raid commercial fruit orchards, but also plays an important role in forest ecology pollinating forest trees and dispersing fruits. Ranges through Myanmar, Thailand, southern Vietnam, Cambodia, peninsular Malaysia, Sumatra, Borneo and Java. The similar *P. giganteus* is found from north-eastern Myanmar through India. The smaller Island Flying Fox, *P. hypomelanus* is found on many offshore islands. All *Pteropus* have declined severely through habitat loss and hunting and some Pacific island species have been hunted to extinction. All species are now protected from international trade.

Flying fox, Pteropus sp.

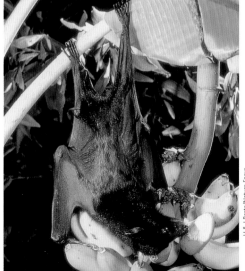

H & J Beste/Nature Focus

ROUSETTES *Rousettus* spp. FA 75–89mm

R. leschenaultii (top);
R. amplexicaudatus (middle);
R. spinalatus (left)

Medium-sized fruit bats, weighing 60–80g, with a relatively narrow muzzle and short tail. Fur short and greyish or brown. Mature males have pale yellow or orange tufts of longer fur on shoulders. Usually roost in caves, and are the only fruit bats with echolocation. Produce clearly audible clicking sounds with their tongues that help them to find their way to roosts in dark caves. Feed on fruits and flower nectar. Geoffrey's Rousette *R. amplexicaudatus* is widespread throughout South-east Asia as far east as Sulawesi. Leschenault's Rousette *R. leschenaultii* occurs throughout mainland South-east Asia. Bare-backed Rousette *R. spinalatus* is known only from Sumatra and Borneo.

SHORT-NOSED FRUIT BATS *Cynopterus* spp. FA 55–80mm

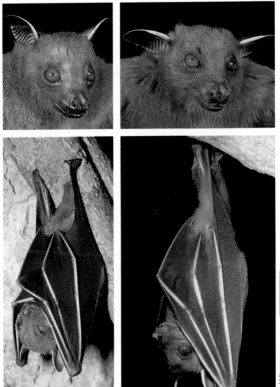

C. brachyotis   C. sphinx

Charles M Francis

Commonest and most widespread fruit bats in region, found in forests from lowlands to hills. Grey to brown, with orange or yellow collar that is darker orange and brighter in adult males. Rims of ears and wing bones whitish. Weights range from 21g in small *C. brachyotis* to 70g in large *C. horsfieldi*. Roost in hollow trees, dense clumps of leaves, or near cave entrances. Sometimes use a 'tent' made by biting the ribs of leaves to make them droop. Feed on fruit and nectar. Small fruits are swallowed whole. Large fruits are squeezed in the mouth to suck out the juices, then the seeds and fibres are spat out. Important dispersers of many pioneer forest trees, thus aiding forest regeneration after disturbance. Single young carried by female for first few days of life. Breeds throughout year in Malaysia, mainly when food is most abundant. Taxonomists recognise five to seven species, differing in size and tooth shape. Up to three species may be found together in all countries within the region.

TAILLESS FRUIT BATS *Megaerops* spp. FA 45–60mm

M. ecaudatus (left);
M. niphanae (below);
M. wetmorei (bottom)

Charles M Francis

These bats have sandy grey-brown or reddish-brown fur with pale grey bases. The ears have dark rims, and the tail is extremely short or lacking. Adult males have a tuft of longer fur on each side of the neck, which varies from pale grey in most species to a striking white in White-collared Fruit Bat *M. wetmorei*. The nostrils are relatively large and protruding, especially in *M. ecaudatus*, which may help the bats to breathe while feeding on soft fruits. Weights range from 14g in *M. wetmorei* to 38g in a large *M. ecaudatus*. In penisular Malaysia, the main breeding season appears to be early in the year, when they give birth to a single young. In Laos, they give birth at the end of the dry season in April and May. The young are carried by the mother when she flies out to forage during first few days of their life. Found in forests from lowlands to 3000m. Appear to be more closely tied to pristine forest than *Cynopterus*. Forage mainly in forest canopy, but may come low to ground along forest edges and streams. Four species are found in region, one or two of which are found in most areas.

DAYAK FRUIT BAT *Dyacopterus spadiceus* FA 75–83mm

Medium-large fruit bat, weighing 75–100g, with short grey-brown fur and heavy-set, strong jaws. Adult males have dull orange-yellow tufts on sides of neck. Massive teeth suggest they can feed on large, hard fruit, but they also eat figs and other soft fruit. Forage mainly in canopy of mature lowland rainforest.

Male (above); female (below)

Have been found roosting in hollow trees. This was the first species of mammal for which males have been found lactating (producing milk). Seem to be at least partly nomadic, moving to new areas when fruit becomes available there. Found in peninsular Malaysia, Sumatra, Borneo and the Philippines.

SPOTTED-WINGED FRUIT BAT *Balionycteris maculata*
FA 40–45mm

Charles M Francis

One of the smallest fruit bats in the region, weighing only 10–15g. Upperparts dark blackish-brown or grey, underparts paler. Wing membranes, finger joints, face and ears have distinctive pale spots. No visible tail and only one pair of lower incisors. Forages in understorey of lowland rainforest, feeding on forest fruits. Roosts in small groups in crowns of palms, in clumps of epiphytic ferns, in hollowed-out termite nests in trees, or rarely in caves. The spotted pattern may provide camouflage when roosting. In lowland forest, breeding has been recorded throughout the year. Range includes peninsular Thailand, Malaysia and Borneo.

BLACK-CAPPED FRUIT BAT *Chironax melanocephalus*
FA 42–46mm

Dark grey or black head, dark grey or brown back, and pale brownish-grey underparts. Chin yellowish, and adults have yellow-orange tufts on sides of neck. No visible tail. Two pairs of lower incisors. Weight varies from 14–19g. Found in forest from lowlands to hills, most often reported above 600m, but may be overlooked in lowland forest as often forages in canopy. Roosts of two to eight have been found underneath epiphytic ferns in Java. Feed on fruits including wild figs. Breeds in February–April in Malaysia, and usually gives birth to a single young. Occurs in peninsular Thailand, Malaysia, Java, Sumatra, Borneo and Sulawesi.

LONG-TONGUED NECTAR BAT *Macroglossus minimus*
FA 38–42mm

Narrow muzzle with tiny peg-like teeth except for needle-like canines. Long tongue with brush-like tip for feeding on nectar and pollen of mangrove, banana, mango and other fruit trees. Roosts under branches and in rolled-up banana leaves. Often solitary, but may roost in groups of five to ten individuals. Occurs in coastal Vietnam and Cambodia, peninsular Thailand, Malaysia, Borneo, Java, the Philippines and islands as far east as Australia. Similar but larger *M. sobrinus* (FA 42–48, weight 18–23g) ranges from eastern India throughout the mainland to Sumatra and Java.

CAVE NECTAR BAT *Eonycteris spelaea* FA 62–70mm

Charles M Francis

L Bruce Kekule

This species has a narrow muzzle and elongated sticky tongue for feeding on nectar and pollen. Distinguished from similar rousettes by lack of a claw on second digit. Has a distinct tail and weighs 45–60g. Usually roosts in caves, often in colonies numbering thousands. Feeds mainly on nectar and pollen, and is a major pollinator of forest trees including commercially important species such as durian, kapok and mangroves. May travel 30km each way from roosting to foraging sites. In peninsular Malaysia, breeds throughout the year. Found throughout South-east Asia, including Sulawesi and the Philippines. Greater Nectar Bat *E. major* (FA 71–80, weight 90g) is darker with more elongate muzzle, and found in northern Borneo and the Philippines.

SHEATH-TAILED BATS *Emballonura* spp. FA 41–48mm

E. alecto (top); E. monticola (bottom)

When disturbed at roost in fallen trees, rock crevices or caves, these bats take up a characteristic pose, holding on by their wrists. Often give high-pitched call before flying off. Distinguished by short tail protruding from tail membrane, small size (weight 4–7g) and uniformly dark-brown fur. Large eyes for an insectivorous bat give excellent vision and may help avoid predators. Often roost in forest understorey, but probably catch insects in middle storey or over canopy. Lesser Sheath-tailed Bat *E. monticola* occurs in peninsular Thailand and Malaysia, Sumatra, Java and Borneo. Greater Sheath-tailed Bat *E. alecto* ranges east from Borneo through the Philippines, Sulawesi and the Moluccas.

BLACK-BEARDED TOMB BAT *Taphozous melanopogon*
FA 60–63mm

Like *Emballonura*, Tomb Bats have short tails that protrude from the membrane and large eyes, but are larger (weights 20–50g) with thicker muzzles. Named after habit of roosting in ancient Egyptian tombs, but also use buildings, rock crevices and caves. Black-bearded Tomb Bat varies from pale buff to dark brown with black chin in males. Pouched Tomb Bat *T. saccolaimus* (FA 71–78mm) is dark brown with white spots above and pure white below. Three other species occur in region, with two

Charles M Francis

to four in each country. Often leave roosts before dark, and can be identified by strong rapid flight, and long narrow wings, often translucent white. Echolocation calls may be audible as high-pitched click. Forage high above canopy for insects.

LESSER MOUSE-TAILED BAT *Rhinopoma hardwickii*
FA 53–64mm

Distinguished by long thin tail, same length as body. Fur soft and short; ears joined across top of head; small ridge on muzzle like a rudimentary noseleaf. Often in relatively hot, dry habitats. In northern areas, may migrate to winter roosts to hibernate. Roost sites include caves, tunnels, buildings and rock crevices. Extremely agile, and can run quickly within roosts or on the ground. Forage in open areas or above forest, catching insects. In India, give birth to single young once a year. Occur from Africa through India with possible records from Myanmar and Thailand. The larger *R. microphyllum* (FA 62–75) occurs from Africa through India and in northern Sumatra.

Harrison Institute

KITTI'S HOG-NOSED BAT *Craseonycteris thonglongyai*
FA 22–26 mm

Not discovered until 1973, this sole representative of the family Craseonycteridae, is one of the world's smallest mammals, weighing only 2.5g. Has small, pig-like nose and lacks any external tail. Eyes relatively small, but ears large. Colour varies from red-brown to greyish. Known only from caves in western Thailand and adjacent Myanmar, where it roosts in colonies, sometimes of several hundreds. The single young is born late in the dry season, and is left at the roost when mother goes out foraging. Leaves cave at dusk and feeds on insects caught in flight around the edges of trees. It is threatened by forest clearance and disturbance in caves.

L Bruce Kukule

MALAYAN SLIT-FACED BAT *Nycteris tragata* FA 50–55mm

Members of family Nycteridae have large ears and a noseleaf consisting of rounded flaps on either side of a narrow slit that extends backwards from nostrils to a hollow on forehead. Tail longer than body, ending in a 'T'-shaped cartilage. Found in tall lowland forest where roosts singly or in small groups. Broad wings and large tail membrane allow it to fly slowly and glean prey from branches or the ground. Feeds on moths, crickets, other insects, scorpions and spiders. Normally has single young. Found in peninsular Myanmar, Thailand, Malaysia, Sumatra and Borneo. A similar species *N. javanica* occurs in Java and is sometimes considered conspecific.

Charles M Francis

FALSE VAMPIRES *Megaderma* spp. FA 54–70mm

M. Spasma (above & below)   *M. lyra. (below)*

Charles M Francis

The family Megadermatidae is distinguished by large ears joined across forehead, an elongated tragus divided into two unequal lobes, a prominent noseleaf, and a large interfemoral membrane joining the legs without any skeletal tail. Greater False Vampire *M. lyra* (FA 67–70, weight 40–50g) is larger, has a small intermediate noseleaf, a tall, parallel-sided posterior noseleaf and a more protruding chin. Lesser False Vampire *M. spasma* (FA 54–61, weight 20–30g) is smaller with a broad intermediate noseleaf and convex sides to the posterior noseleaf. Despite their name, they do not drink blood, unlike the true vampire bats found in the Americas. Instead, they feed by gleaning prey from branches or the ground. They often rely on prey-generated sounds such as rustling of dead leaves, mating calls of crickets, or possibly echolocation calls of other bats, to locate prey. The smaller *M. spasma* feeds mainly on arthropods, but has been known to prey upon small bats in traps. It is found throughout the region. The larger *M. lyra* has been reported eating a wide variety of vertebrates including mice, small bats, frogs and birds, as well as arthropods. It is similarly widespread on the mainland, but has not been found on the islands.

HORSESHOE BATS *Rhinolophus* spp. FA 37–72mm

(Clockwise from below)
R. macrotis; R. paradoxolophus;
R. luctus; R. trifoliatus; R. sedulus

Charles M Francis

A broad, horseshoe-shaped anterior noseleaf, a tall pointed posterior noseleaf, large ears without a tragus, and a medium-length tail fully enclosed in the interfemoral membrane characterize the family Rhinolophidae. All members of the family are currently placed in the single genus, *Rhinolophus*, which has nearly 60 species worldwide including about 30 species in the region. There is tremendous diversity among species in the size and shape of the noseleaf. Some species have extra flaps on the base of the noseleaf, others have tufts of hair on the back. The echolocation calls have a distinctive structure, with a long constant frequency component that allows them to detect moving prey through Doppler shift (the same principle used to measure the speed of moving objects with radar). A moth flapping its wings will cause the frequency of the echo to vary, allowing the bat to distinguish the insect from nearby leaves. All of these calls are well above the range of human hearing, ranging from 23kHz to over 100kHz.

30

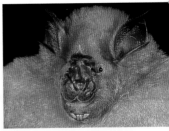

R. coelophyllus (top left); R. acuminatus (top right); R. malayanus (above); R. affinis (below)

Each species has a distinctive call frequency, to which it is finely adapted in its internal hearing morphology as well as body size and noseleaf shape. The noseleaf appears to help focus the calls, which are emitted through the nostrils. The reason for such elaborate flaps on the noseleaf is unknown, but species with low frequency calls (longer wavelengths) tend to have a larger noseleaf and ears. Up to 10 species may be found together in a particular patch of forest, specializing in different insect prey according to their echolocation calls and body size (high frequency calls are better for detecting small insects). Most species roost in caves, often in large colonies of several hundred up to more than a hundred thousand. They will also sometimes roost in small colonies in hollow trees or rock crevices. A few species may roost in trees among dead leaves, including *R. trifoliatus* whose pale brown and yellow fur may provide camouflage. Most other species are shades of dark brown or grey, but many species have two colour phases, with some individuals bright orange or rufous. Horseshoe bats usually give birth to a single young which clings to the mother during the day, but is left hanging in the roost when she flies out to forage at night. At least a few species are found in all parts of the region.

Charles M Francis

ROUNDLEAF BATS *Hipposideros* spp. FA 34–100mm

H. pomona (above)

The family Hipposideridae includes nine genera worldwide, of which four occur in the region. The most diverse genus is *Hipposideros* with over 25 species in the region. Like horseshoe bats, roundleaf bats have a distinctive noseleaf with a horseshoe-shaped anterior leaf, but the posterior leaf is generally rounded and there is no protruding sella. The noseleaf varies from small and simple in species such as *H. pomona* to large and elaborate with multiple leaflets in species such as *H. armiger*.

H. cineraceus (above); H. galeritus (below left); H. rotalis (below right)

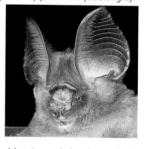

Most species can be distinguished by size and the shape of the noseleaf, but recent studies suggest that some cryptic species appear nearly identical externally, but differ in genetics and echolocation calls. Several new species have been discovered recently, including *H. rotalis* which was first described in 1999 from the Annamite mountains in Laos. The echolocation calls are similar in structure to those of *Rhinolophus*, but are generally shorter in duration, and for

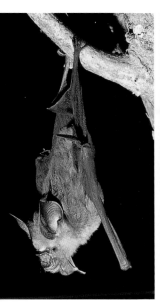

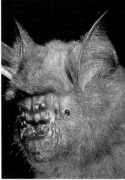

Charles M Francis

(Clockwise from above)
H. larvatus; H. armiger; H. diadema
male; H. diadema female and young

a given body size are about twice the frequency, ranging from 50kHz to nearly 200kHz. High frequency calls may be helpful for catching moths, because many moths have ears that enable them to hear lower frequency bat calls. Males of many roundleaf bats have a small sac behind the noseleaf that emits a strong smelling substance that may play a role in communication or sexual selection. Roundleaf bats usually roost in caves, sometimes in very large colonies, although some species also roost in smaller groups in hollow trees. Females give birth to a single young, usually once a year, though populations in some areas of Malaysia, where the seasons are less strongly marked, appear to have multiple peaks of breeding. At least one species, and usually several, are found in all parts of the region.

TAILLESS ROUNDLEAF BATS *Coelops* spp. FA 34–43mm

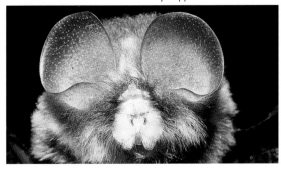

A poorly known group, with few records. Distinguished by well-developed interfemoral membrane with no visible skeletal tail, a noseleaf with large flaps on each side and small funnel-shaped ears. Known to roost in caves and hollow trees. Echolocation calls very high-pitched, suggesting a small insect diet. Smaller *C. robinsoni* (FA 34–37) has broadly rounded anterior lobes to the noseleaf, while larger *C. frithi* (FA 40–43) has narrow lobes. Both are reported from scattered localities throughout the region. The closely related *Paracoelops megalotis* also lacks a tail, but has larger ears and a different noseleaf. It is known only from one specimen collected in Vietnam in 1945.

TRIDENT ROUNDLEAF BAT *Aselliscus stoliczkanus* FA 41–44mm

Charles M Francis

Distinguished from other roundleaf bats by the posterior noseleaf which is raised into three points. The tip of the tail protrudes from the interfemoral membrane. The fur varies from pale orange-brown to dark grey-brown with whitish bases. Young are greyer, as in many other species of bats. Males may have two tufts of long white hairs on their chests, each with a clump of a musky-smelling yellow secretion that probably is used for communication or mate selection. Has usually been found roosting in limestone caves. Ranges from southern China throughout mainland South-east Asia, but not recorded from the islands.

PIPISTRELLES *Pipistrellus* spp. FA 26–42mm

(Clockwise from above)
P. paterculus; P. paterculus;
P. coromandra; A. cuprosus

Charles M Francis

Pipistrelles are in the world-wide family Vespertilionidae, which is distinguished by a simple nose without a noseleaf, separate ears with a distinct tragus, and a long tail enclosed within the interfemoral membrane. Bats traditionally assigned to this genus are distinguished by having two upper premolars, but recent research indicates they belong in several different genera. Thick-thumbed pipistrelles (genus *Glischropus*) are similar but with thickened thumb pads and the second upper incisor displaced outwards. Great Evening Bat *Ia io* is much larger (FA 75–80, weight 50g). About 23 species of pipistrelles are currently recognized from the region, but many are difficult to identify and their ranges are still poorly known. Roost in hollow trees, among leaves, in rock crevices, buildings or less frequently in caves. During the breeding season, some species form separate maternity colonies for the females and young (like many other bats), while others roost in small groups with a single male and several females. Most species usually give birth to twins, though single young and more rarely triplets have been reported. Breeding appears to be year-round in some populations, but strongly seasonal in others. Pipistrelles are adapted for foraging in open areas, such as clearings or over the forest canopy, where they catch insects. Most parts of South-east Asia support at least a few species in this group.

M. formosus (above & left);
M. muricola (below left);
M. siligorensis (below right)

Charles M Francis

Myotis is the most widespread genus of bats, reported from every continent except Antarctica, with over 100 species around the world. At least 23 species are found in South-east Asia, ranging in size from the Small-toothed Myotis *M. siligorensis* weighing 2.3–2.8g to the Chinese Myotis *M. chinensis* which may weigh 30–40g. The genus is distinguished by having three premolars above and below (usually) and by tall distinctly shaped ears that are narrow near the top and have a long tragus inside that tapers to the tip and is usually bent forwards. Most species catch insects in flight over streams, fields or forest. A few species have enlarged feet, such as the Rickett's Myotis *M. ricketti*, and feed by scooping insects or small fish off the surface of the water with their feet. Fish are located by using sonar to detect the

M. annectans (above)
M. ricketti (right & below)

Charles M Francis

M. horsfieldii (above left & right)

ripples they make when they break the surface. Echolocation calls, like those of most Vespertilionidae, are frequency-modulated, sweeping through a range of frequencies from high to low, but all inaudible to humans. Myotis roost in a variety of locations including caves, rock crevices, hollow trees, and even rolled-up young banana leaves. Most species give birth to a single young each year which may be carried by the mother when it is still small. Several species occur in all countries within the region.

FALSE-SEROTINES *Hesperoptenus* spp. FA 24–43mm

Charles M Francis

H. tickelli   *H. blanfordi*

False-serotines are distinguished from pipistrelles mainly by dental characters: they have only one upper and lower premolar on each side, and the second upper incisor is displaced inwards so that the canine and first incisor are touching, or nearly so. Other genera in the region with only a single premolar include *Philetor* and *Eptesicus*, which are both distinguished by skull characters. Four species are currently known from the region. Least False-serotine *H. blanfordi* has a very short forearm (FA 24–27), but is relatively heavy, weighing 6–7g. Tickell's False-serotine *H. tickelli* (FA 55–60, weight 14–21g) has distinctive yellow-brown fur and wings. Tome's False-serotine *H. tomesi* (FA 50–53, weight 30–32g) and Doria's False-serotine *H. doriae* (FA 38–41) are reddish-brown or dark brown. False-serotines forage in relatively open areas such as over streams or forest canopy. Presumably roost mainly in hollow trees, but few roosting sites have been found. At least one species has been reported from most parts of mainland South-east Asia, as well as Borneo.

BAMBOO BATS *Tylonycteris* spp. FA 22–32mm

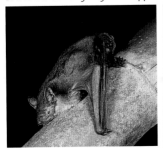

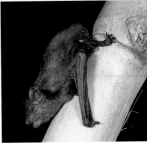

T. robustula *T. pachypus*

Bamboo bats have flattened skulls and flexible flat bodies, allowing them to squeeze into narrow slits to roost inside bamboos. Lesser Bamboo Bat *T. pachypus* (FA 22–28, weight 3.5–5g) can fit through a slit 3.9mm wide. Greater Bamboo Bat (FA 27–32, weight 7–10g) requires a slit 4.5mm wide. Thickened pads on thumbs and ankles enable these bats to grip on bamboo. Species are distinguished by size, by sleek dark-brown fur on Greater, and fluffy reddish-brown fur on Lesser. Roost in groups of one adult male with up to fifteen adult females, or small groups of males. Give birth to two young. Feed over forest on flying insects, with preference for swarming termites. Can fly independently at six weeks old. Both species occur throughout region. Disk-footed Bat *Eudiscopus denticulus* of Laos, Thailand and Vietnam also has flattened skull and large pads on feet, but has different dentition and ear shape.

HARLEQUIN BAT *Scotomanes ornatus* FA 55–62mm

Distinctive and colourful species readily distinguished by dark orange-brown fur with white spots on the shoulders and sides, and white stripe down middle of back. Has only a single pair of upper incisors and upper premolars. Weight 22–30g. Roosts in small trees, hanging from branches among clumps of leaves where coloration may give some camouflage. Also recorded around limestone caves. Feeds on insects caught in flight. Ranges from southern China south to central Myanmar, Thailand, Laos, Vietnam and Cambodia. Asian house bats *Scotophilus* spp,. found throughout the region, have similar dentition, but plain yellow-brown fur.

Charles M Francis

BENT-WINGED BATS *Miniopterus* spp. FA 34–53mm

M. australis (above);
M. magnater (below)

Bent-winged bats (also known as long-fingered bats) have the last bone of the longest finger approximately three times as long as next to last bone. Widely distributed Old World genus, with 11 or more species worldwide, and at least five species in the region. All species appear very similar except for differences in measurements; populations of the same species living in different areas may differ in size, making taxonomy confused. Roost in caves forming colonies of up to 100,000 bats. Females in maternity colonies give birth to single young after gestation of up to five months. Young may be left in a 'crèche' with thousands of young bats per square metre. Amazingly, mothers are able to find their own young on returning from feeding. Young are independent at about two months. Found throughout the region with two to five species occurring together.

WOOLLY BATS or BUTTERFLY BATS *Kerivoula* spp. FA 26–49mm

Charles M Francis

K. hardwickii

These small bats range in size from the 2g Least Woolly Bat *K. minuta* to the 13g Papillose Woolly Bat *K. papillosa*. Recognized by funnel-shaped ears with a long narrow tapered tragus and three well-developed upper and lower premolars. Long woolly fur makes some species resemble a ball of orange fluff with wings. Eleven species are currently recognized in the region. The Painted Bat *K. picta*, has bright orange fur and striking black and orange wings. The

K. minuta

Clear-winged Woolly Bat *K. pellucida* has long ears, pale brown fur with whitish bases and translucent brown wings. Some species pairs, such as Least and Small Woolly Bats *K. intermedia* with short ears and orange-brown fur with dark bases, can only be distinguished by skull measurements. All woolly bats have large wings relative to body weight, enabling them to fly slowly and manoeuvre tightly among trees in the forest. Their echolocation calls are very high frequency compared with other similar-sized bats, suggesting they often feed by gleaning small insects from leaves or branches. They roost among dead leaves or in hollow trees, and give birth to a single young. This genus is found throughout the region, with the highest known diversity in Borneo where eight species occur.

K. pellucida (above); K. picta (below)

Charles M Francis

GROOVE-TOOTHED BATS *Phoniscus* spp. FA 31–39mm

P. jagorii (above); P. atrox (below)

Closely allied to woolly bats *Kerivoula* spp., but differ in white, unpigmented tragus with notch at base, groove on outer surface of upper canines, and colourful fur that is dark at base, with a pale then a dark band, then golden or orange tips. The two species differ mainly in size: Gilded Groove-toothed Bat *P. atrox* (FA 31–35) weighs 3.5–5 g, and Frosted Groove-toothed Bat *P. jagorii* (FA 36–39) weighs 7–9g. Thought to roost and forage in similar manner to *Kerivoula* spp. Gives birth to single young. Smaller species reported from peninsular Thailand and Malaysia, Sumatra and Borneo, while larger species known from south China and scattered localities throughout region, as well as Philippines.

TUBE-NOSED BATS *Murina* spp. FA 28–41mm

M. aurata

M. aenea

Charles M Francis

The tubular nostrils on these bats superficially resemble those of many fruit bats, but as far as is known, *Murina* feed mainly on insects and other arthropods. The function of their elongated nostrils remains a mystery. The ears are rounded, with a long pointed tragus. The dentition is unusual among insectivorous bats in having relatively large premolars, with reduced molars. At least 11 species occur in the region, differing in coloration and dental characters. Several species, apparently not closely related, have body hairs with alternating bands of pale and dark fur ending in shiny golden tips. Other species have simple coloration with orange or brown fur with dark grey bases or, in one species, white bases. In most species, the interfemoral membrane and the bases of the wings are also covered in hairs. Like the woolly bats *Kerivoula*, these species have high frequency echolocation calls and appear to feed by gleaning insects off branches. They roost in trees and vegetation, often hidden within clumps of dead or dry leaves. Little is known of their breeding, but they have been reported giving birth to both single young and twins. Tube-nosed bats are found throughout the region, with the highest known diversity in Laos where seven species have been reported, though most of these species probably also occur in neighbouring countries.

M. cyclotis

HAIRY-WINGED BAT *Harpiocephalus harpia* FA 48–52 mm

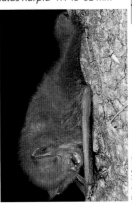

Charles M Francis

Closely related to tube-nosed bats, but larger (weight 15–20g) with more massive teeth, greater reductions of the premolars, and extensive long orange hairs over interfemoral membranes and bases of the wings. Fur varies from rufous-brown with grey bases to bright orange. Females substantially larger than males with more massive teeth and jaws. Roost sites remain undescribed. Their massive teeth would be appropriate for feeding on hard-bodied insects such as beetles. These bats are rarely captured, so their range is poorly known, but at least one species has been reported from most parts of the region.

43

WRINKLE-LIPPED BAT *Tadarida plicata* FA 40–44mm

Charles M Francis

Bats of the family Molossidae, also known as free-tailed bats, are distinguished by a medium-length fleshy tail that protrudes for at least half its length beyond the end of the interfemoral membrane. The ears are thick and joined over the top of the head by a band of skin. Wrinkle-lipped Bat has a heavily wrinkled upper lip. These bats may roost in huge colonies of up to a million or more in large caves or sometimes in buildings. The evening departure from these roosts is one of the most impressive wildlife sights in South-east Asia as huge dark clouds of bats spiral into the sky and disperse for tens of kilometres in all directions to feed on aerial insects such as termites, ants and moths. Many predators such as falcons, bat hawks, eagles, owls and even hornbills, are attracted to these colonies to capture bats. A more serious threat may be disturbance or exploitation by humans. In Laos, where bats are regularly eaten, several thousand bats may be captured in one evening, which could eventually wipe out the colony. This species occurs throughout South-east Asia. Three other species of *Tadarida* have more restricted ranges in the region and are distinguished by size, dental characters, and the shape of the ears. The Naked Bat *Cheiromeles torquatus* (FA 74–83mm) is in the same family, but is one of the heaviest insectivorous bat in the world, weighing 150–200g. Largely naked except for a few scattered hairs, it roosts in colonies up to several thousand individuals in hollow trees or caves. It is known from peninsular Malaysia, Sumatra, Java and Borneo.

PRIMATES Order Primates

South-east Asia has a high diversity of primates, including two prosimians, one great ape, several species of gibbons, macaques, langurs, proboscis monkeys, doucs and snub-nosed monkeys. Several species are threatened or endangered because of habitat loss or persecution.

SUNDA SLOW LORIS *Nycticebus coucang* TL 30–38cm

Gerald Cubitt

This small tail-less primate may be seen most readily by shining a light through trees at night and looking for reflective eyeshine. The fur is mainly brown to grey-brown, paler underneath, with a pale streak between the eyes, dark marks around the eyes, and a bold black stripe down the back. Adults weigh 1–2kg. Lorises walk slowly and deliberately through trees, but can move quickly when necessary, such as when catching prey. Diet comprises insects, birds, small mammals and lizards, as well as fruit, nectar and pollen. Lorises are active at night, sleeping in tree forks or in clumps of branches during the day. Mainly solitary, but occasionally seen in pairs or family groups. Give birth to single young (sometimes twins) which remain with mother for six to nine months. Found throughout mainland South-east Asia and on Sumatra and Borneo, but not Java. Pygmy Loris *N. pygmaeus* which is similar in shape but much smaller (250–400g), more orange in colour and lacks a dorsal stripe, is found in parts of Laos, Vietnam and Cambodia. Some authorities consider form north of peninsular Thailand to be a district species, Asian Slow Loris *N. bengalensis*, differing in paler head with less distinct stripe pattern.

WESTERN TARSIER *Tarsius bancanus* TL 30–38cm

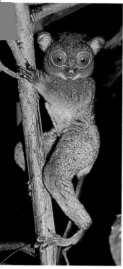

WWF Malaysia/MPS

Tarsiers are easily recognized by their large, forward-facing eyes, long hind legs, and long naked tail with a tuft of hairs at the end. The slender toes and fingers have enlarged pads at the tips. Tarsiers travel by jumping from tree to tree through the forest understorey, much like a tree frog. Able to jump up to 6m horizontally, though an average jump is about 1.4m. Largely nocturnal, the tarsier can rotate the head to see directly behind it. It also has acute hearing which may help in the location of large insects or small vertebrate prey. Tarsiers may be found in primary or secondary forest, where there are high densities of saplings and prey. A largely solitary species, with a home range of about 2–3 hectares. Females give birth to a single young after a six month gestation period. The young is born fully furred, can leap after about a month, and starts to catch its own food shortly afterwards, but takes 15–18 months to mature. Western Tarsier is found only in southern Sumatra, Borneo and a few small offshore islands. Three other species occur on Sulawesi, and one in the Philippines.

Alain Compost

LONG-TAILED MACAQUE *Macaca fascicularis* TL 92–109cm

L. Bruce Kekule

Macaques can be distinguished from langurs by distinctive head shape, more pinkish faces and more muscular bodies. Long-tailed Macaque has the longest tail of any Asian macaque, similar in length to its head and body. Body fur varies from greyish-brown to reddish. In Thailand, these macaques are most common in coastal areas and along large rivers, but farther south they are found in a range of habitats including hill forest, lowland forest, plantations and secondary forest. Diet is omnivorous, including invertebrates such as shellfish or crabs, as well as nestlings, small mammals, fruits and leaves. May become a pest, raiding rice crops, fruit orchards or vegetable gardens, and entering towns or even houses to scrounge food. Highly gregarious, being found in troops of up to 70 individuals. Gives birth to a single young after gestation of five or six months. Range is generally south of Rhesus Macaque *M. mulatta*, in southern Thailand, Laos, Vietnam, Cambodia, peninsular Malaysia, Sumatra, Java, Borneo and the Philippines.

WWF Malaysia/Gerald Cubitt

RHESUS MACAQUE *Macaca mulatta* TL 67–86cm

Indraneil Das

Rhesus Macaque is similar in body size (3–6kg) to Long-tailed Macaque *M. fascicularis* but can be distinguished by its shorter tail (about 40–60% of head and body length), and strong reddish tinge to hind quarters. The bare skin around the face is light pink. This gregarious animal occurs in large troops of 50 or more, and is often found around villages, towns or temples. Like other macaques, this species can climb well, but often forages on the ground, escaping from disturbance by running through undergrowth or climbing trees. Diet is a variety of plant and animal foods, but near towns will also eat garbage and hand-outs. Ranges from Nepal through northern India, across southern China and in central and northern Myanmar, Thailand, Laos, Vietnam and Cambodia.

ASSAMESE MACAQUE *Macaca assamensis* TL 70–100cm

This species has a similar length tail to Rhesus Macaque *M. mulatta*, about half its head and body length, but is larger and heavier (5–10kg), has longer hair, grey instead of reddish hind quarters, and a darker brown face. An upland species, usually found from 500–3500m. Appears to feed frequently in trees, but will drop to the ground, then run off through the undergrowth if disturbed. Diet consists of fruit, buds, insects and other small animals such as lizards. Found from eastern India through Myanmar, Thailand (south to the northern peninsula), northern Laos and Vietnam.

Thomas Geissmann

SOUTHERN PIG-TAILED MACAQUE *Macaca nemestrina*
TL 60–80cm

Adult male (above); immature (below)

This macaque is distinguished by its large muscular build (adult males may weigh 11–13.5kg, although females are only half the size), and short thin tail about one third head and body length, that is carried sticking up over the back. The body fur varies from greyish-brown to reddish, with a distinctive cap of short dark hairs on crown, and a fringe of long hairs around face. The facial skin is pinkish-brown. A more nomadic species than other macaques, often travelling through the forest in a large troop, and not reappearing for several months. The omnivorous diet includes fruits, seeds, buds, insects and other small animals. Captive animals can be trained to climb coconut trees and pick the fruits. Occurs in peninsular Thailand, Malaysia, Sumatra and Borneo. Replaced by closely related *M. leonina* in rest of mainland south-east Asia.

STUMP-TAILED MACAQUE *Macaca arctoides* TL 50–71cm

WWF Malaysia/Dino Sharma

Stump-tailed Macaques (average weight 8–12kg) have a similarly heavy build to Pig-tailed Macaque *M. nemestrina*, but have longer shaggier fur and an extremely short tail (less than 10% of head and body length). The bare facial skin, which extends to

the forehead, and bare skin on rump, is reddish-brown and may turn bright red at times. Found mainly in the uplands, in primary or secondary forest to 2000m, including limestone outcrops. Spends much of time on ground, feeding on seeds, buds, insects and small animals, but will climb trees to pick fruit or shoots. May form large troops of up to 50 individuals, and has been known to raid crops. Ranges from India through southern China and through most of mainland South-east Asia to northern peninsular Malaysia, but generally rare through range, and may be declining.

Thomas Geissmann

WHITE-THIGHED LANGUR *Presbytis siamensis* TL 120–140cm

Langurs can be distinguished from macaques by longer tails (usually much longer than head and body), generally greyer faces (often with pale eye rings), lighter build, and distinctive head shapes, with hairs on crown often raised into a crest. The White-thighed Langur is variable in colour, ranging from pale grey to mid-brown or dark brown. The underparts are usually much paler. Newborn young of Presbytis langurs have pale grey, almost whitish fur, with dark limbs, a dark crown and a dark stripe down back. This species has a distinctive call, a harsh rattle, followed by a loud 'chak-chak-chak-chak'. Groups of up to 25 individuals have been reported, but five or six

Thomas Geissmann

is more common. Like other langurs, diet consists mainly of young shoots and leaves, although may also eat fruit and seeds. Found in peninsular Malaysia and Sumatra. Closely related Banded Langur *P. femoralis* has dark outer thighs and occurs in peninsular Thailand, Malaysia and Sumatra.

MITRED LANGUR *Presbytis melalophos* TL 120–140cm

Mitred Langur is sometimes considered part of the same species as Banded Langur *P. femoralis*. The back varies from red-brown to orange or grey, with the feet and tail the same colour (instead of black in Banded Langur). Underparts are white or buff and cheeks pale. Occurs in most of Sumatra except in the east where it is replaced by Banded Langur. Several closely-related species occur in the region, although in most areas no more than one or two species occur together. North Sumatran Langur *P. thomasi* with a white forehead, is restricted to northern Sumatra. Mentawai Langur *P. potenziani* is found only on the

Thomas Geissmann

Mentawai Islands. Grizzled Langur *P. comata* is the only species on Java. Hose's Langur *P. hosei*, with white marks around the face is found in northern Borneo, and White-fronted Langur *P. frontata* with a white spot on the forehead, occurs in south Borneo. Maroon Langur *P. rubicunda*, found in much of Borneo, is uniform reddish-orange with blue-grey facial skin. May occur in same forest as Hose's Langur.

SUNDAIC SILVERED LANGUR *Trachypithecus cristatus*
TL 100–130cm

Alain Compost

Sundaic Silvered Langur is dark grey all over, with pale tips to fur giving silvery appearance. Facial skin is generally dark grey. Infants are a contrasting bright orange. In some parts of Sabah, reddish adults have also been seen. In most parts of range it is found mainly in coastal areas, especially mangroves, and in forests along large rivers, although may occur in inland forests in Indochina. Diet consists of young leaves, shoots, fruits and flowers. Troops average five or six individuals in Sarawak, but groups of 10 or more are commonly encountered in peninsular Malaysia, with groups of up to 40 occasionally reported. Young recorded throughout year, suggesting no marked breeding season. Largely arboreal, this species is found in peninsular Myanmar, southern Thailand, Malaysia, Sumatra and Borneo. Replaced in Indochina by *T. margarita*. In Java, Ebony Langur *T. auratus* has two colour phases, black or reddish-orange, and occupies similar coastal habitats.

Male (left)

WWF Malaysia/Gerald Cubitt

Mother and young

DUSKY LANGUR *Trachypithecus obscurus* TL 110–115cm

Gerald Cubitt

Indraneil Das

Most individuals are dark grey, with paler grey on hind legs and crown, and a pale grey or whitish belly patch, sharply demarcated from darker back. The bare skin on the face is dark grey with bold white interrupted rings around the eyes and a white patch over the mouth. Newborn young are bright golden-yellow, like other *Trachypithecus*. Call is a loud, double snort or grunt, rendered as 'cengkong' in Malay. Occurs in a wide range of habitats from montane to coastal forests including some offshore islands. Feed mainly on leaves, buds and fruits. In peninsular Malaysia, tends to feed higher and in larger trees than Banded Langur *Presbytis femoralis*, when both species occur together. Found in troops of about 15 individuals with a single male. Gives birth to a single young after a gestation period of about five months. Range includes southern Myanmar and Thailand, peninsular Malaysia and some adjacent islands.

PHAYRE'S LANGUR *Trachypithecus phayrei* TL 110–150cm

This langur is highly variable in colour, ranging from nearly black through rusty-brown to grey, with usually paler underparts. Hind legs and tail are usually the same colour as back. Like Dusky Langur *T. obscurus*, has whitish rings around eyes and a white patch on both lips. Generally found in upland areas, away from humans, in dense forest and bamboo. In some parts of Thailand occurs on limestone cliffs, and may use caves. In parts of Myanmar, this species was hunted for its large gallstones ('bezoar stones') which developed as a result of drinking calcium-rich spring water, and which are highly prized in Chinese medicine. Populations in these areas have largely been wiped out by hunting. Found in northern Myanmar, Thailand, Laos, Vietnam and southern China.

L Bruce Kekule

FRANCOIS'S LANGUR *Trachypithecus francoisi* TL 130–155cm

Francois's Langurs are mainly black, both above and below, with white stripes on cheeks from mouth to ears. Several similar, closely related species occur in region. Hatinh Langur, *T. hatinhensis* has a longer moustache. Lao Langur, *T. laotum* has a pale forehead. Black Langur, *T. ebenus* has an all black head. Cat Ba Langur, *T. poliocephalus* has a pale yellow head and dark brown body. Delacour's Langur, *T. delacouri* has a large white patch on the rump. All species occur in restricted areas in central Laos, central and northern Vietnam and extreme southern China, mostly around limestone outcrops. Several species are endangered due to habitat loss and excessive hunting.

W Robichaud/WCS

DOUCS *Pygathrix* spp. TL 110–150cm

Red-shanked Douc, *P. nemaeus* is one of the most strikingly coloured mammals in south-east Asia, with large patches of speckled grey, white and black on its body, dark reddish brown legs, a long white tail, bright chestnut below the ears, and a yellowish face fringed with long white hairs. Grey-shanked Douc, *P. cinereus* also has a yellowish face but the body is paler and the legs are grey. Black-shanked Douc, *P. nigripes* differs in having blue-grey facial skin, and black legs. All species are Found in both mature monsoon forest, and well-developed second growth forest, in groups of up to a dozen individuals. Feed mainly on leaves and buds, supplemented with fruit. Crash noisily through trees when relaxed, but can disappear quickly and silently through tree-tops when danger threatens. Found only in central and south Vietnam, Laos and north-east Cambodia with Red-shanked in the north, Grey-shanked in the middle, and Black-shanked to the south. Populations have been substantially reduced by habitat loss and hunting, with Grey-shanked the rarest species.

P. nemaeus

Thomas Geissmann

P. nemaeus (left); *P. nigripes* (right)

Thomas Geissmann

TONKIN SNUB-NOSED MONKEY *Rhinopithecus avunculus*
TL 110–150cm

Xi Zhinong/Nature Picture Library

Tonkin Snub-nosed Monkey is one of the least known and most endangered primates in South-east Asia. Has cream-coloured forehead, face and underparts; brown to black back and outside of legs and arms except for a pale patch on elbow; and a long dark tail with a pale tip. Facial skin is blue-grey, except for pink lips. Nose is flattened. Body fur is relatively long. Closely related to Golden Monkey *R. roxellana* of China, which is one of the hardiest monkeys, surviving cold, snowy winters. Found only in two small reserves in the mountains of northern Vietnam, in forests around limestone. Feeds on fruit, leaves and seeds. Lives in groups of one male with several females and young. Most of original habitat has been lost, and no more than a few hundred individuals remain.

PROBOSCIS MONKEY *Nasalis larvatus* TL 120–150cm

Adult males (16–22kg) have a distinctive and greatly enlarged nose, while females (only 7–11kg) have a smaller upturned nose. Body fur of both sexes is mainly orange, darker above and paler below, with pale grey limbs and white rump and tail. The function of the enlarged nose is unclear, but may be related to sexual selection by females.

Alain Compost

Male (above); female with young (below)

The main social structure is a harem, consisting of one mature male and several females and their young. Subadult males may travel in bachelor groups until they grow large enough to acquire their own harem. Proboscis monkeys eat leaves, hard fruits and seeds. Their habitat is largely restricted to coastal mangroves, and riverine forest, although they may travel up to 2 km away from river each day, in search of tender, young leaves. They have partially webbed hind feet, and can swim readily. Found only in Borneo, along coasts and large rivers, where their population appears to have declined considerably in recent years due to habitat loss. Closely related Pig-tailed Langur or Simakobu *Simias concolor*, found only on the Mentawai Islands, has upturned nose like juvenile Proboscis Monkey, but fur is either all dark brown or all golden.

WHITE-HANDED GIBBON *Hylobates lar* TL 45–50cm

WWF Malaysia/Slim Sreedharan

Gerald Cubitt

Gibbons are smaller and less closely related to humans than other apes. All gibbons have long arms, lack a tail, and are highly arboreal. They normally swing by their arms from branch to branch through the trees, but occasionally can be seen walking on their hind legs along a branch or on the ground. White-handed Gibbon varies from almost black through dark brown to pale buff or cream-coloured, but can always be distinguished by its white or buff hands, feet and face marks. The colour variation is not related to sex, and animals of either colour can be found together in the same family group. Gibbons are most readily located by their loud, ringing calls which can be heard for up to 2km. The great call of female White-handed Gibbon is a long wail, rising in irregular swoops through two octaves. Gibbons feed on a variety of fruits, young leaves, shoots and flowers, along with occasional insects and other animal food. Females usually give birth to single young, after a gestation period of about seven months. The young takes six to eight years to reach maturity, but the mother may have another young after two to four years. This species is found in southwest Yunnan (China), southern and eastern Myanmar, Thailand, north-west Laos, peninsular Malaysia and Sumatra.

PILEATED GIBBON *Hylobates pileatus* TL 45–50cm

Pileated Gibbons are sexually dimorphic in colour. Adult males are all black, except for white border around face, while adult females are pale grey-brown with black patches on breast and crown. Young animals of both sexes are greyish-white, acquiring black patches like adult females at four to six months. This species co-occurs with White-handed Gibbon *H. lar* in some parts of central Thailand, and occasional hybrids have been recorded, but the females have quite distinct calls, and normally reproduce only within their own species. The great call of the female accelerates to a series of rapid bubbling notes, unlike the much slower, more soaring notes of White-handed Gibbon. Found in dry or moist evergreen forests, or semi-deciduous forest that have not been excessively disturbed. Eats fruit

Gerald Cubitt

and leaves. Groups usually consist of a pair with their offspring, varying in size from two to six individuals. Limited to south-eastern Thailand, western Cambodia and south-west Laos, west of the Mekong River.

AGILE GIBBON *Hylobates agilis* TL 45–50cm

Agile Gibbons are quite variable in colour, ranging from cream to dark brown, generally with white eyebrows and sometimes with pale cheek patches that may join under the chin. Hands and feet are same colour as rest of fur, unlike White-handed Gibbon *H. lar*. Bornean race is less variable than others, being usually buff below and dark brown above. Great call of female is similar to that of White-handed Gibbon, with rising and falling notes, but also has diagnostic 'whoo-aa' calls. Both calls are quite different from Bornean Gibbon *H. muelleri*. Agile Gibbon has three distinct populations, each with limited overlap with other gibbon species. One population occurs in peninsular Malaysia and Thailand, with White-handed Gibbon to both north and south. The population in southern Sumatra

Thomas Geissmann

is replaced to the north by White-handed Gibbon. In Borneo, *H. a. albibarbis* is found in west and central Kalimantan between the Kapuas and Barito Rivers, being replaced in rest of island by Bornean Gibbon.

JAVAN GIBBON *Hylobates moloch* TL 45–60cm

Alain Compost

This species has very long, silvery-grey fur, usually with a darker chest and cap, some white around face, and dark hands. Lives in family groups that jointly defend a territory of approximately 17ha. Restricted to tall forest, feeding on fruit, supplemented by leaves and some insects. This is the only species of gibbon in its restricted range in western Java, where it is suffering from severe habitat loss and excessive trapping for the pet trade. It is considered endangered.

BORNEAN GIBBON *Hylobates muelleri* TL 45–60cm

Alain Compost

The fur is generally grey-brown with a darker chest and cap, but there is some variation in colour, with some individuals nearly black, and others very pale. The female great call is distinctive, consisting of a series of loud, bubbly whoops that carry for up to 2 km. Normally found only in tall forest, but can tolerate some selective logging provided that sufficient tall, fruit-bearing trees, are retained. Eats ripe fleshy fruits, young leaves and small insects. Occurs in small family groups consisting of a breeding pair with up to three young, which defend a territory of 20–30ha. Does not reach sexual maturity until six to eight years of age. Gives birth to one young at approximately two yearly intervals. Restricted to Borneo, where it is the only species in most of island except in south-west, between Kapuas and Barito rivers, where replaced by Agile Gibbon *H. agilis*.

KLOSS'S GIBBON *Hylobates klossi* TL 45–50cm

This gibbon has uniformly glossy black fur, without any pale markings on face or hands. Fur is much sparser than in other gibbons, with only a third as many hairs. Similar in weight to other small gibbons (5.5–6.5kg), and has similar-sized territories of 30–35ha. Unlike other gibbons, males and females rarely duet, but both sexes sing solos which are among the most elaborate and complex of gibbon songs. Male song bouts may last for up to 40 minutes, often pre-dawn. Eats mainly fruit, supplemented by insects and small amounts of leaves; eats relatively more insects and fewer leaves than other gibbons. Restricted to the four Mentawai Islands of Siberut, Sipura, north and south Pagai, off the west coast of Sumatra, where it is endangered by habitat loss.

Alain Compost

HOOLOCK GIBBON *Hoolock hoolock* TL 45–60cm

Males are mainly black, sometimes with brown tinge, while females are pale brown with darker fur on belly, throat and sides of head, and pale cream cap. Both sexes have white eyebrows and female has thin white line around face and around each eye. Slightly larger (average weight 6.8kg) than other small gibbons. Both sexes have small throat sac, but not nearly as well-developed as in Siamang *S. syndactylus*. Restricted to tropical evergreen and semi-deciduous forest, where has a home range of 20–30ha. Eats mainly fruit and leaves with small amounts of insects. Lives in small family groups of pair and offspring. Song usually given only as a duet, with sexes giving alternating high and low notes. Found only in northern Myanmar and adjacent parts of Assam (India), Bangladesh, and China. Much of its original habitat has been lost, and it is considered endangered in most of range.

Thomas Geissmann

BLACK CRESTED GIBBON *Nomascus concolor* TL 45–55cm

Male (above); female (right)

Males are black with small crest of erect hairs on top of head and no pale facial markings. Females yellowish to pale grey-brown, with contrasting dark belly, throat, face, cap, fingers and toes. Males have small throat pouch used to produce distinctive 'boom' call as part of song. Both sexes sing duets, female starting with accelerating series of notes, male ending with series of up and down notes. Found only in northern Vietnam, south Yunnan and Hainan Island in China.

WHITE-CHEEKED GIBBON *Nomascus leucogenys* TL 45–55cm

Female (above); male (right)

Adult male black with distinct crest, and contrasting large white patches on each cheek. Females pale to golden-yellow without any dark hairs on belly, with long, rounded dark crown patch, and usually white brow band and pale white rim around face. Lives in small family groups consisting of pair and offspring. Found in tropical evergreen or semi-evergreen forest. Calls similar to Black Crested Gibbon *N. concolor*, but female great call is faster with about twice as many notes. Restricted to north Laos, north Vietnam and southern China, where much of original habitat now lost. Replaced in central Laos and Vietnam by similar *N. siki*. Both species are endangered.

BUFF-CHEEKED GIBBON *Nomascus gabriellae* TL 45–55cm

Male (above); female (right)

Similar to White-cheeked Gibbon *N. leucogenys*, but male has orange-brown patch on chest, and cheek patches are pale creamy-orange or yellow. Females orange with small black cap and black tufts on ears. Songs and calls resemble Black Crested Gibbon N. concolor. This species probably lives in family groups, like other gibbons, and feeds on fruit and leaves. Restricted to southern Vietnam, east Cambodia and probably Laos where it is endangered by habitat loss and hunting.

SIAMANG *Symphalangus syndactylus* TL 75–90cm

The largest gibbon, completely black with somewhat shaggy fur, and a bare throat pouch that inflates when the animal is calling. Most readily located by great call, which includes resonant booms and whoops, leading up to a series of loud barks and a piercing yell. Under ideal conditions, the call may carry for several kilometres. Found mainly in tall forest in hilly areas. Family group consists of two adults and up to three dependent young, but unmated individuals are solitary. Gives birth to one young approximately every two or three years. Diet consists mainly of leaves, fruits (especially figs), flowers and shoots. It eats relatively more leaves than smaller gibbons. Restricted to peninsular Malaysia and Sumatra.

Gerald Cubitt

63

ORANGUTAN *Pongo pygmaeus* TL 120–150cm

Alain Compost

The Orangutan is the only great ape in South-east Asia, and is generally considered less closely related to humans than the chimpanzees of Africa. Readily distinguished by large size, orange body fur without a tail, and usually slow and deliberate movements. Mature male develops large fleshy flaps on sides of face, longer darker fur, and may weigh 50–100kg. Adult female is generally smaller, weighing 35–50kg. These highly intelligent apes remain with their mother for between five and seven years after birth, learning about the types of fruits in the territory. Female reaches adult body size at about eight years old, but male may not reach maturity until 15 years. These apes reproduce quite slowly, giving birth only every four or five years. The only primate in the region to make a nest, consisting of broken twigs and

Sub-adult male

Alain Compost

small branches woven together, usually high in a tree. Diet consists mainly of fruits and young leaves, and Orangutans roam over areas of 100 hectares or more in search of food, returning to individual trees throughout the year when the fruits start to ripen. Most of the year adults remain apart, but several individuals may converge on a single, large fruiting tree. Adult male can produce a loud roar which may carry for over a kilometre. Orangutans are found only on Sumatra and Borneo. The Sumatran form is genetically distinct, with longer face, lighter longer hair, and male develops a long beard; some authorities have suggested this may be a distinct species, *P. abelii*. Populations in both areas are endangered by habitat loss and excessive hunting, including trapping for the pet trade.

CARNIVORES Order Carnivora

About 56 species of carnivores occur in the region, including ten cats, four dogs, two bears, and a variety of weasels, martens, badgers, otters, civets and mongooses. Despite the name, some species, such as the civets, eat mainly fruit and insects, although they will eat meat if the opportunity arises. Many of the species are seriously threatened by excessive hunting and habitat loss.

GOLDEN JACKAL *Canis aureus* TL 80–100cm

Gerald Cubitt

This jackal is smaller than the Dhole *Cuon alpinus*, with greyish-brown fur, sometimes tinged yellowish or reddish, and black-tipped fur on shoulders and lower back that may form a pale saddle-like pattern over the shoulders. The tail has a black tip. Prefers relatively open country. Mainly nocturnal, although also active during the day where few people are around. Usually travels alone or in small family parties of a pair with dependent young. Omnivorous, eating a variety of plant and animal foods. Prey includes small mammals, birds, amphibians and reptiles. Will also feed on carrion, including remains of prey killed by other larger predators. May enter villages or camps to scavenge garbage, catch rodents or even raid crops such as sugar cane. Breeds at any time of year, and usually give birth to four or five young after a gestation period of two months. Both male and female help to feed young. Usually remain within a territory of two or three square kilometres. A widespread jackal occurring in north and east Africa, southern Europe, Pakistan and India through Myanmar and northern and western Thailand.

DHOLE *Cuon alpinus* TL 130–160cm

The Dhole is the largest wild dog in the region. The fur colour is variable, but usually dark reddish-brown above, whitish below, with a long, bushy black tail. The ears are relatively short and rounded. These dogs live in a wide range of forested habitats, from dense scrub to tall lowland forest to montane forest. Found in packs of five to twelve, usually consisting of a breeding pair and their offspring. Occupy dens which they either make themselves or take over from another animal. The usual litter size is four to six. Hunt in packs, often attacking animals much larger than themselves, such as deer, pigs or Gaur but also capture smaller prey such as hares, rodents or insects. They may eat carrion, and packs of Dholes have been known to chase larger carnivores such as Leopards or Tigers away from prey. Ranges from India to China and Siberia and south throughout most of mainland South-east Asia, as well as Sumatra and Java, but not Borneo. Declined throughout range due to persecution and habitat loss.

ASIAN BLACK BEAR *Ursus thibetanus* TL 130–190cm

WWF Malaysia/Azwad Mohd Noor

This bear has shaggy black fur with a white crescent on the chest and large rounded ears. A few individuals may be brown or reddish-brown. Considerably larger than Sun Bear *U. malayanus*, averaging 110–150kg (male) and 65–90kg (female). Frequents moist deciduous forests and scrubby areas, most often in the hills, sometimes up to 3600m. Solitary, except during the breeding season, and most active at night. During the day, sleeps in tree hollows, caves or rock crevices. Main diet is fruits, invertebrates and small mammals, but has been reported to hunt large mammals up to the size of a buffalo. Females give birth to two young after a gestation period of seven or eight months. The young may remain with the female up to three years, sometimes after a second litter has been born. Ranges from Pakistan through the Himalayas to Siberia, and in northern Myanmar, Thailand, Cambodia, Laos and Vietnam. Has suffered considerably from habitat loss and human persecution, including the trade in bear gall bladders and bile, and is now rare in many areas.

SUN BEAR *Ursus malayanus* TL 100–150cm

Gerald Cubitt

Sun Bear is the smallest bear in the world, weighing 27–65kg. The fur is short and sleek, all black except for white on the face and a white U-shaped mark on the chest. The ears are short and rounded. They inhabit dense forests at all elevations, and are excellent climbers. Although they may be seen during the day, they are most active at night, spending the day sleeping in trees. Diet includes a wide variety of invertebrates, fruits and plant material including the hearts of coconut palms. They capture termites by tearing open the mounds with their sharp claws, and also catch small rodents and birds such as Red Junglefowl *Gallus gallus*. Litters contain one or two young, and captive animals have been recorded producing two litters per year. Normally, young remain with the female until full grown. Found in forested areas throughout South-east Asia, including Sumatra and Borneo, but not Java.

YELLOW-THROATED MARTEN *Martes flavigula* TL 80–100cm

Indraneil Das

Long, slender mammal with sleek brown fur; black feet, tail and sides of neck; and a pale to bright yellow upper chest and throat. Considerable geographic variation: in northern Thailand has nearly black crown and pale belly, while in Malaysia, including Borneo, has brown crown and dark brown belly. Martens are excellent climbers, but also spend much time on ground. Varied diet includes fruits, insects, eggs and nectar, as well as frogs, birds or squirrels. Inhabits wide range of forest types, from lowlands to high mountains. Most active during day. Found throughout South-east Asia, and as far north as Siberia and as far west as Pakistan.

HOG BADGER *Arctonyx collaris* TL 75–120cm

L Bruce Kekule

Thickset, large badger with pig-like nose. Varies from yellowish to greyish with dark legs and pale tail. Usually shows black stripe through eye and another on chin, and pale stripe from forehead to nose. Sleeps during day in rock crevices or burrows that it digs for itself, using its extremely long, white claws. Active by night, foraging for tubers, roots, insects, worms and other animals that it grubs out of the ground, probably finding them by smell. Its lower incisors are angled forwards, to help with digging. Litter size two to four. Generally found in forested areas to 3500m, throughout China, Myanmar, Thailand, Laos, Cambodia, Vietnam and Sumatra.

FERRET-BADGERS *Melogale* spp. TL 40–48cm

Anthony Lamb

Kinabalu Ferret-badger (above); Small-toothed Ferret-badger (below)

Rod Williams/Bruce Coleman Collection

Ferret-badgers vary from brown to grey-brown with a distinctive black and white face pattern, and tail about half length of body. Found in forested areas or grasslands. Mainly nocturnal, sleeping in burrows or crevices during day. Varied diet of insects, worms, other invertebrates and fruit. They grub much of their food from the ground, but can also climb trees. Female gives birth to one to three young in a burrow. Taxonomists recognize up to four species in the region. All are similar in colour, and differ mainly in size and shape of the teeth. Kinabalu Ferret-badger *M. everetti* occurs in hill forest between 1000–3000m on Mount Kinabalu in Borneo. Javan Ferret-badger *M. orientalis* is found on both Java and Bali. Large-toothed Ferret-badger *M. personata* occurs from Nepal through Thailand and Vietnam, while Small-toothed Ferret-badger *M. moschata* occurs from Vietnam north through China.

ORIENTAL SMALL-CLAWED OTTER *Aonyx cinerea* TL 65–90cm

Smallest otter, weighing less than 3kg. Distinguished from other otters by small size and short claws that do not extend beyond the tips of the toes. Mainly grey-brown with a white throat. Generally found in family groups of four to six but sometimes up to 12 individuals. Occurs in a wide variety of habitats containing both permanent water and some tree cover, from coastal estuaries and large rivers to small hill streams and ponds. Relies more heavily on molluscs and crustaceans than other otters, although also feeds on fish. Litter size one or two, and may breed twice a year. Distributed throughout South-east Asia, being commonest otter in most areas.

SMOOTH OTTER *Lutrogale perspicillata* TL 105–120cm

WWF Malaysia/Oon Swee Hock

Distinguished from other large otters in South-east Asia by its smooth coat, naked nose and more flattened tail. Considerably larger than Oriental Small-clawed Otter *A. cinerea*, weighing 7–11kg. Found mainly in lowlands and coastal areas, including estuaries, river mouths, reservoirs, lakes and streams. Usually found near water, but may travel long distances overland to new water bodies. Often found in family groups. Litter size one or two, in dens dug into the banks of rivers. Eats fish, turtles, crustaceans and shellfish. In India, they have sometimes been trained to chase fish into nets. Occurs throughout South-east Asia, but has declined in many areas from loss of wetland habitats.

EURASIAN OTTER *Lutra lutra* TL 90–120cm

Gerald Cubitt

This otter has a rough coat with a dark brown back, paler underparts, less flattened tail and naked nose pad. Feeds mainly on fish, but also eats other aquatic animals. Builds den at edge of water. More montane than other Asian otters, but also sometimes found in lowlands. Occurs through much of Europe and northern Asia, as well as in scattered locations in northern Myanmar, Thailand, Laos, Vietnam and Sumatra. Hairy-nosed Otter *L. sumatrana* is similar but nose pad is hairy, and underparts almost as dark as upperparts except for white chin. It is recorded from southern Vietnam, Cambodia, peninsular Thailand, Malaysia, Sumatra and Borneo.

LARGE INDIAN CIVET *Viverra zibetha* TL 120–130cm

L Bruce Kekule

Large civet with black and white throat markings, mottled black and brown spots on grey-brown coat, and five or six complete white bands alternating with black bands on tail. A crest of black erectile hairs extends from back of neck to base of tail. Like most civets, largely nocturnal, sleeping during day in dense bushes. Forages on ground for lizards, small mammals and insects; also eats roots, tubers and fruits including oil palm nuts. Litter size two to four, and may have two litters per year. Occurs from India through southern China and all of mainland South-east Asia, but not Indonesia. Large Spotted Civet *V. megaspila* of mainland South-east Asia is similar, but has heavier spots on flanks, and tail shows incomplete pale rings at base, with distal half all-black.

MALAY CIVET *Viverra tangalunga* TL 95–115cm

Gerald Cubitt

Coat colour varies from grey and black to brown. Distinguished from Large Indian Civet by smaller size, and by tail pattern with many more rings (10–15 black ones) connected along the top by a black stripe. Diet includes invertebrates, small vertebrates and plant matter, largely taken on forest floor. May enter forest camps or villages looking for food scraps or garbage. Occurs mainly in forested areas, or nearby cultivated lands, from lowlands to hills. The only species of *Viverra* civet in Sumatra, Borneo, and the Philippines, though its range overlaps both Large Indian Civet and Large Spotted Civet in peninsular Malaysia.

LITTLE CIVET *Viverricula indica* TL 84–106cm

Differs from larger cousins by neat rows of spots along its sides, lack of black crest along back, and evenly spaced complete dark and light bands around pale-tipped tail. Most frequently encountered in open grassy areas, or scrub, rarely in thick forest. Hunts mainly at night, resting under rocks or in tall grass or bushes during day. Eats a wide variety of animals including birds, small mammals, lizards, insects and plant matter, and may enter villages to scavenge for food. Litter of four to five young in a burrow under rock or base of tree. Extends from India to China, south throughout mainland South-east Asia, as well as Sumatra and Java.

BANDED LINSANG *Prionodon linsang* TL 65–77cm

Boldly-patterned small civet with white or buff fur with large black or dark brown spots and bars on upperparts. Legs short, long bushy tail has complete black and pale rings. Spotted Linsang *P. pardicolor* differs in having smaller discrete spots, not large splotches, on back. Unlike most other civets, linsangs have retractile claws (like cats). Active at night, sleeping during day in hole in ground or in tree. Forages on ground and in trees, hunting a variety of small mammals, birds, reptiles, amphibians and invertebrates. Largely restricted to forest and adjacent areas. Litter size two or three. Occurs in southern Myanmar and Thailand, peninsular Malaysia, Sumatra, Java and Borneo. Spotted Linsang found from Nepal through northern Myanmar, Thailand, Cambodia, Laos and Vietnam.

COMMON PALM CIVET *Paradoxurus hermaphroditus*
TL 85–140cm

Alain Compost

A widespread civet, varying from olive-grey to cream, with three dark stripes on back and additional dark spots on flanks, sometimes forming indistinct lines. Usually has a dark 'mask' highlighted by paler fur on forehead and behind cheeks, and sometimes with a pale spot below eye. Nocturnal and usually solitary. Often seen on the ground, but feeds mainly in trees, where it eats fruits and animals. May be an important seed disperser for various forest tree species. Litter size three, born in a den in a hollow tree or under a boulder. Found from India through southern China, throughout mainland South-east Asia, Sumatra, Java and Borneo. Also occurs on the Philippines, Sulawesi and many eastern Indonesian islands where it may have been introduced.

SMALL-TOOTHED PALM CIVET *Arctogalidia trivirgata*
TL 90–120cm

Neil Lucas/Nature Picture Library

More arboreal than Common Palm Civet *Paradoxurus hermaphroditus*. Varies from light grey-brown to dark grey, but always has three thin black stripes on back, with no spots on flanks. Individuals in Thailand have thin white stripe on nose; in Borneo, the head is usually mostly black. Tail longer than head and body, partly prehensile, and can be wrapped around branches when climbing. Largely nocturnal, feeding mainly on fruit, but also eats insects, small mammals, birds, lizards, nectar and pollen from flowers of forest trees. Extends from eastern India through Myanmar, Thailand, Laos, Vietnam, Cambodia, peninsular Malaysia, Sumatra, Java and Borneo. Prefers tall forest, and greatly reduced in many areas through habitat loss.

MASKED PALM CIVET *Paguma larvata* TL 100–140cm

WWF Malaysia/Slim Sreedharan

Colour variable, from blonde to dark brown. Lacks stripes or spots, but dark marks around eyes, highlighted by pale cheeks and forehead, give a mask-like pattern. Long tail usually black, at least distal half, but sometimes with white tip. Nocturnal, resting in hollow cavities or forks in trees during the day. Largely arboreal, but also hunts on ground. Feeds on fruits, leaves and flowers, and occasionally small mammals or birds. Like other civets, produces strong musky secretion from glands at base of tail. Largely solitary. Litter size one to four, reaching adult size in three months. Occurs from northern India through southern China and south through mainland South-east Asia, Sumatra and Java.

BANDED CIVET *Hemigalus derbyanus* TL 70–82cm

Alain Compost

Pale brown to orange-brown with black bands across upperparts; black and white stripes on face. No spots on flanks. Tail black except for pale bands at base. Frequently forages on the ground, but climbs well. Feeds on insects and other invertebrates, or on aquatic animals captured along streams. Found only in peninsular Thailand, Malaysia, Sumatra and Borneo. Owston's Civet *H. owstoni*, in southern China, Laos and northern Vietnam, is similar but has extensive spotting on legs. Hose's Civet *Diplogale hosei*, from mountains of northern Borneo, is similar in shape but upperparts and legs are uniformly dark grey or grey-brown, belly and throat white.

BINTURONG *Arctictis binturong* TL 110–180cm

Indraneil Das

Gerald Cubitt

This animal, sometimes known as Bear-Cat, is the largest of the civets, weighing up to 20kg. It has long, shaggy dark fur, with tufts on its ears, and a long prehensile tail. Many of the hairs are tipped white, giving a frosted appearance. Nocturnal, and mainly arboreal, it is a slow, deliberate climber, using its tail to hold onto branches as it climbs. The tail can grip strongly enough to support the animal completely if necessary. Most individuals probably feed mainly on fruits, especially ripe figs, but also eat a variety of animals. Females may breed at most times of year, sometimes producing two litters per year. The average litter size is two, but ranges from one to six. Young reach adult size in about a year, and can breed shortly thereafter. Found in tall forests through most of South-east Asia, including Sumatra, Java and Borneo.

CRAB-EATING MONGOOSE *Herpestes urva* TL 70–80cm

L. Bruce Kekule

Mongooses are closely related to civets, but are largely diurnal, with small ears, a distinctive posture, and a bushy tail that tapers to the tip. Crab-eating Mongoose is the largest species in the region (3–4kg), distinguished by its tall stance, shaggy brown to grey-brown fur, with a long pale streak on the shoulder. Glands by the tail secrete a strong-smelling musky fluid. Forages mainly along streams in dry forests and open areas. Readily enters water to capture aquatic animals such as crabs, frogs, fish and molluscs. Found from eastern India through south China, and mainland South-east Asia into northern peninsular Malaysia.

SMALL ASIAN MONGOOSE *Herpestes javanicus* TL 60–70cm

Alain Compost

Smaller (0.5–1.0kg) than Crab-eating Mongoose *H. urva*, with relatively shorter legs, speckled dark brown or reddish brown fur and no shoulder stripe. Lives in open country, such as scrub, rice fields or dry forests, feeding mainly on rats, as well as birds, eggs, insects and reptiles including cobras. Breeds throughout the year, producing 2–4 young at intervals as short as 4 months apart. Found from Iran through India, mainland South-east Asia and Java. Three other mongooses occur in dense lowland rainforests in the region. Short-tailed Mongoose *H. brachyura* is dark brown, speckled orange, with pale chin and relatively short bushy tail, and found in peninsular Malaysia, Sumatra, Borneo and Palawan. Collared Mongoose *H. semitorquatus*, larger and paler with a pale band under the neck and a longer tail, is confined to Sumatra and Borneo. Hose's Mongoose *H. hosei* is known from only a single specimen from Borneo.

TIGER *Panthera tigris* TL 240–290cm

Alain Compost

Sumatran race

Tigers are readily distinguished by their large size and the pattern of vertical black stripes on orange fur over most of the body. Populations from the northern parts of the range are the largest living cats, weighing up to 300kg, but southern populations, such as those on Sumatra, are much smaller, with males weighing 100–140kg, and females 75–110kg. Tigers inhabit a wide range of forested habitats, from dry savannah to wet tropical rainforest. Unlike most cats, they often swim or play in water. Diet consists mainly of large mammals, including pigs, deer and wild cattle, though they will feed on almost any available animal food. Most tigers avoid people, and are difficult to find, but a few individuals

Indochinese race

Tesny Whittacker/FLPA

Tiger pug marks

Gerald Cubitt

learn to prey upon humans and become dangerous. Tigers are usually solitary animals, coming together only to mate. Litter size is normally one to four. In historical times, tigers ranged from the Indian subcontinent throughout much of South-east Asia as far south as Java and Bali (but not on Borneo) and as far north as Siberia. However, populations on Bali, Java and Singapore are now extirpated, and populations elsewhere are greatly reduced and threatened with extinction. The species is listed by IUCN as endangered, with major threats including loss of habitat, reduced prey populations, and illegal hunting, trapping, and trade, especially for alleged medicinal properties.

LEOPARD *Panthera pardus* TL 180–220cm

Alain Compost

L Bruce Kekule

The second largest cat in the region, males average 55kg and females 30kg. Pale colour phase has black spots clustered into rosettes on a pale buff coat. Dark colour phase (sometimes called Black Panther) appears all black, but the rosettes are still discernible in good light. Spotted phase is more common in northern part of region, while black phase is more common in south, though both forms can interbreed, producing some offspring of each colour. Largely nocturnal in South-east Asia but may be diurnal elsewhere. Found in a wide variety of habitats, especially forested areas, and often climb trees. Feed on many types of animals, including pigs,

Black form

deer, monkeys, large birds, rodents or even insects. Usually avoids humans, but some individuals become bold and prey on domestic animals. Solitary except during the breeding season. Produces litter of from one to four cubs which remain with mother until full grown at 1.5–2 years. Leopard is the most widespread of the world's large cats, ranging throughout Africa and most of southern Asia, including all of mainland South-east Asia as well as Java, though not apparently in Sumatra or Borneo. However, it has declined throughout its range due to persecution and loss of habitat.

CLOUDED LEOPARD *Neofelis nebulosa* TL 120–175cm

Alain Compost

Gerald Cubitt

Although nearly as long as Leopard *Panthera pardus*, Clouded Leopard is substantially smaller, weighing only 16–23kg. The fur has a distinctive pattern with large rosettes rimmed in black and separated by pale reticulations forming 'clouds'. Found in a wide variety of forests from lowlands to hills up to 2500m. With its long tail, this cat is a good climber, and often sleeps in trees, but recent studies suggests it hunts mainly on the ground. Sometimes waits on a low branch to pounce on prey. Active by day or night, and captures a variety of animals including birds, monkeys, pigs, deer and sometimes young cattle or buffalo. Little is known about their breeding in the wild, but captive animals produce litters of one to five young, usually two. The species occurs from Nepal through eastern China, and through most of mainland South-east Asia including Sumatra and Borneo. Populations have apparently declined considerably in many areas from loss of habitat combined with excessive hunting for their attractive pelts.

MARBLED CAT *Pardofelis marmorata* TL 90–110cm

WWF Malaysia/Ken Scriven

Resembles a much smaller version of Clouded Leopard *Neofelis nebulosa*, weighing only 2–5kg, having similar cloud-like patterns on fur, but these are generally less distinct, giving a more marbled pattern, rather than distinct 'clouds.' Shows more small black spots on legs, and even young animals have thin stripes from eye to crown, instead of a row of spots like Clouded Leopard. A poorly known species, probably due to its highly arboreal and nocturnal habits. Thought to prey largely on birds and small rodents such as squirrels or rats. Apparently rare, reported from scattered localities in mainland South-east Asia, as well as Sumatra and Borneo.

ASIAN GOLDEN CAT *Catopuma temminckii* TL 120–130cm

L Bruce Kekule

A large cat, weighing 12–15kg. Varies from golden-brown to greyish-brown, occasionally dark. Generally lacks stripes or spots, but has distinctive pale lines, often bordered black, running across cheeks and from inner corners of eyes to crown. Underside of tail white, contrasting with upperside. Northern individuals sometimes heavily spotted on flanks. Found in range of forest types from open deciduous to tall rainforest. Active by day or night, feeding on rodents, hares, small deer, birds and lizards. Reported hunting in pairs; male may stay with female when she has young. Litter size one or two in a den in hollow tree or among rocks. Occurs from eastern India through southern China, mainland South-east Asia and Sumatra. Bay Cat *C. badia*, found on Borneo, is similar, with dark and light colour phases, but is one of the most rarely seen cats.

FISHING CAT *Prionailurus viverrinus* TL 97–107cm

Alain Compost

Gerald Cubitt

Pale grey or olive-brown, with numerous small black spots in rows along sides and back, forming stripes on back of neck. Spots are generally smaller than in Leopard Cat *P. bengalensis*, and it is twice as heavy, weighing 7–11kg. Found in brushy or scrubby habitats, usually hunting near water, feeding on fish, crabs or frogs, as well as small rodents and birds. Fishes from rock or ledge overhanging water, and scooping out small fish with its paws. Gives birth to two or three young, which may reach full size in eight months. Found from India, through most of South-east Asia including Sumatra and Java, Myanmar, central and northern Thailand, Indochina, but not peninsular Malaysia or Borneo.

LEOPARD CAT *Prionailurus bengalensis* TL 67–84cm

Gerald Cubitt

The most frequently encountered wild cat in most of South-east Asia. Distinguished by yellow-brown fur with large black spots 2.5cm or more in diameter, some stretching into stripes, especially on shoulders and back of neck. Occurs in wide range of habitats from tall forest to scrub or plantations, and sometimes even villages or suburban areas. Mainly nocturnal, hunting small vertebrates including frogs, lizards, birds and small mammals. May enter caves to feed on fallen bats, or rodents. Litter size one to four; young may be sexually mature within a year. Ranges from Nepal and northern India through eastern China to Siberia, and throughout mainland Southeast Asia including Sumatra, Java and Borneo. Swims well, and has colonized many offshore islands.

A Lamb

87

FLAT-HEADED CAT *Prionailurus planiceps* TL 59–65cm

Gerald Cubitt

Alain Compost

This small cat, about size of domestic cat (weight 1.5–2kg), is distinguished by short tail (about 25% of head and body length), small rounded ears and flat forehead. Varies from dark brown to brownish-grey, without spots or stripes. Throat and chin white, with two dark stripes from eye to ear on each side of face, and pale stripe above each eye to forehead. Has short claws which cannot be retracted. Mainly nocturnal, hunting for frogs, crustaceans and fish along small forest streams, and readily entering water to catch or wash food. Its behaviour is little known. Found only in peninsular Malaysia and adjacent Thailand, Sumatra and Borneo.

JUNGLE CAT *Felis chaus* TL 75–95cm

JP Zwaenepoel/Bruce Coleman Collection

Gunter Ziesler/Bruce Coleman Collection

Relatively long-legged, medium-sized cat (4–6kg), with plain ashy grey to yellowish-brown body fur, dark stripes on legs and tail, and black tufts on tips of each ear. Despite name, lives in open grassy habitats, deciduous forests, scrub, and along streams. Terrestrial, rarely climbing trees. Relatively long legs for chasing after prey. Diet includes rodents, hares, lizards, frogs, birds, small deer. Ranges from Middle East through much of Asia to mainland South-east Asia north of peninsular Thailand. Population has declined in many areas, because preferred habitat is in areas where many people also settle, leading to exposure to hunting.

WHALES AND DOLPHINS Order Cetacea

Although cetaceans resemble fish, they are warm-blooded mammals that breathe air, give birth to live young, and suckle them on milk. They comprise two suborders, the Mysticeti including large whales which feed by filtering plankton and shrimp with baleen, and the Odontoceti including dolphins with teeth. At least 25 species probably occur in the seas around South-east Asia, ranging in size from large baleen whales, exceeding 20m in length, to small dolphins and porpoises which may be less than two metres long. Only one species regularly leaves coastal areas and enters large rivers.

IRRAWADDY DOLPHIN *Orcaella brevirostris* TL 200–250cm

The Irrawaddy Dolphin has a small, rounded dorsal fin, behind the middle of the back, a high rounded forehead with no beak, and broad rounded flippers. It is pale grey to dark bluish-grey, with no distinctive markings. Usually occurs in small groups of up to 10 individuals. Often feeds in murky or silted waters. This species, like bats and many other dolphins, probably relies on echolocation (sonar) to find prey in dark waters. It emits a series of clicks and listens for echoes of those sounds bouncing off prey items. Main diet is fish, along with crayfish, shrimp and other crustaceans. In some areas, it is apparently believed by fishermen to help herd fish into nets. Usually, when these dolphins breathe, the head is seen first, then the back with the low dorsal fin. They rarely jump out of water. Give birth to single young which is about 40% of female body length. Regularly seen in estuaries and bays in coastal areas of much of South-east Asia, but can also live in fresh water, and swims inland up several major rivers, including the Mekong as far inland as Laos, and the Mahakam River in Borneo. Some populations, such as along the Mekong River in Laos, are now severely endangered due to a variety of causes including hunting, accidental mortality in fishing nets, reductions in food supply from over fishing, and disturbance from boats.

SEA COWS Order Sirenia

The sea cows are a small order with four species still surviving, three of which (the manatees) live in tropical seas in the Americas and around Africa, while only the Dugong occurs in South-east Asia. Although fully aquatic, they are not closely related to whales and dolphins, and are thought to have evolved independently from an ancestor related to elephants. Their forelimbs are modified into flippers, and they lack hind limbs.

DUGONG *Dugong dugon* TL 240–270cm

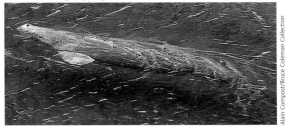

Alain Compost/Bruce Coleman Collection

Duane Yates/ Nature Focus

A slow-moving mammal that spends its entire life in water. Has a large rounded head, with nostrils on top of muzzle (instead of above the eyes), and the mouth underneath; also has a forked tail and flippers like dolphin, but lacks any dorsal fin. The brownish-grey skin is thick and leathery. Normally weighs 230–360kg, but individuals up to 908kg have been recorded. Feeds entirely on aquatic plants, especially sea grasses, which are pulled up with the thick flexible lips and shaken to remove sand and debris before being swallowed. The large, sac-like stomach stores this bulky food. Dives are usually short, lasting up to three minutes. Usually found singly or in small groups, but in Australia large herds of up to 100 animals have been reported. Gives birth to a single young, occasionally twins, after a 13–14 month gestation; the young is suckled for up to 18 months. Dugongs reach sexual maturity at 9–10 years, but may not breed until 15 years. The interval between births is typically three to seven years. Males have enlarged incisors that they apparently use when fighting for mates. Formerly occurred in coastal waters and estuaries throughout South-east Asia, but greatly reduced by hunting, habitat loss and pollution which has killed many of the sea grasses. Now a very scarce species, rarely encountered anywhere in the region.

ELEPHANTS Order Proboscidea

The elephants are a distinctive order with only two living species, one in Africa and one in Asia, though many additional species of mammoths and mastadons are known from the fossil record, some of which survived until the past few thousand years.

ASIAN ELEPHANT *Elephas maximus* SH 250–300cm

Alain Compost

Elephants are easily identified by their large size, thick grey or grey-brown skin, and long trunk. The tusks of females and young males are rarely visible, but some large males have tusks up to one metre long. Asian elephants differ from their African relatives by their smaller size, much smaller ears, and single 'finger' at the tip of the trunk (instead of two). The tracks of adults are 35–50cm across with five toe marks on the front foot, and four on the back. Found in a wide variety of forested areas including monsoon forest, lowland rainforest, swamp and plantations. Active by both day and night, although prefer to remain in shade during heat of the day. Diet comprises a wide variety of plants, including palms, bananas, twigs, bark and leaves from a

Alain Compost

Gerald Cubitt

range of trees and shrubs, and vines. This species may cause extensive damage in plantations of young palms, bananas, sugar cane or other crops. Occurs singly or in small herds usually led by a dominant male. Female gives birth to single young, or rarely twins, after gestation of 21 months. In many areas, elephants have been tamed and used as beasts of burden, including for dragging logs out of the forest during logging operations – a much less damaging process than modern tractors. Formerly widely distributed from India throughout mainland South-east Asia, and on Sumatra and north-eastern Borneo, but greatly reduced by hunting and forest loss.

ODD-TOED UNGULATES Order Perissodactyla

The odd-toed ungulates include the horses, tapirs and rhinoceroses. Although they feed mainly on grasses and leaves, they do not have the elaborate stomachs of the even-toed ungulates. Only one tapir and two species of rhinoceros occur in the region.

ASIAN TAPIR *Tapirus indicus* SH 90–105cm

Adult (top); juvenile (bottom)

A distinctive mammal with black on head and forelegs, white in middle, and black on hind legs. Young are very different, with dark brown fur and longitudinal pale stripes. Nose is elongated into a short trunk, and tail is very short. Tracks are similar to those of rhinos, but only 15–17cm across, with four toe marks in the front feet (fourth only visible in soft ground). Tapirs inhabit dense rain forests, from lowlands to hills, near streams or other permanent water sources. Diet includes a wide range of leaves and other plants in the forest undergrowth and along the edges of clearings and streams. A largely solitary species except when breeding. Gives birth to single young weighing 6–7kg after gestation of just over a year; the young remains with female for six to eight months. Sparsely distributed in southern Thailand, Myanmar, peninsular Malaysia and Sumatra, with numbers having been greatly reduced by forest loss and hunting.

LESSER ONE-HORNED RHINOCEROS *Rhinoceros sondaicus*
SH 140–170cm

Alain Compost

Alain Compost

Distinguished from Asian Two-horned Rhinoceros *Dicerorhinus sumatrensis* by its larger size, and by having three folds of skin across the back, one behind the neck, one behind the shoulders, and one over the rump. The single horn is relatively small, rarely exceeding 15cm. The tracks, which have only three toes, exceed 23cm in diameter in mature adults. Rhinos inhabit dense rainforest, usually in the lowlands, in areas with mud wallows and water, feeding on shoots, twigs, young foliage and fallen fruit. They are largely solitary, coming together only to breed. Give birth to single calf after a gestation period of 16 months; the young may suckle for up to two years. Female probably does not breed more regularly than every four or five years. This is one of the most endangered large mammals in the world, having been nearly wiped out by habitat loss and hunting. Formerly distributed across much of South-east Asia, the only remaining populations now known are at Ujung Kulon National Park in extreme western Java, where perhaps 50 individuals survive, and in southern Vietnam where there have been a few recent reports of surviving individuals at one site only.

ASIAN TWO-HORNED RHINOCEROS
Dicerorhinus sumatrensis SH 100–140cm

WWF Malaysia/Guy Baskin

Alain Compost

Distinctly smaller than Lesser One-horned Rhinoceros *Rhinoceros sondaicus*, with two horns (although these are often small and inconspicuous) and only two folds of skin across the back, one behind the shoulders and one over the rump. The young are very hairy, but the hairs become reduced and less conspicuous as they get older. The tracks have three toes and are 17–22cm across. Formerly occurred in a wide variety of habitats from lowland swamps to hill forest, but most remaining populations are in hillier, more remote areas. Able to climb very steep slopes, despite its large size. The presence of salt licks seems to be an important factor limiting its range. Diet includes fallen fruit, leaves, twigs and bark from a wide variety of plant species. Often wallows in mud pools to cool down and avoid insects. Rhinos are solitary except when breeding, and give birth to a single young, which remains with female for 15–18 months. Probably does not breed until seven or eight years of age. Formerly occurred throughout most of South-east Asia, but now limited to small scattered populations in peninsular Malaysia, Sumatra, Borneo and possibly parts of Thailand, southern Myanmar, Cambodia or Vietnam. This species is highly endangered by habitat loss and hunting.

EVEN-TOED UNGULATES Order Artiodactyla

The even-toed ungulates are represented in South-east Asia by pigs, deer, cattle, goats and antelopes. All but the pigs are ruminants, with complex stomachs to aid with digestion of leaves, grasses and other fibrous plant foods.

EURASIAN WILD PIG *Sus scrofa* SH 60–80cm

Alain Compost

This species differs from other pigs in the region by its dark body, narrow pointed muzzle without warts or bumps, and a mane of black hair extending halfway down back. The young are dark brown or blackish with white stripes along the body. Found in Europe and north Africa as well as most of Asia, the Wild Pig is the ancestor of the modern domestic pig. Feeds on a wide variety of plant and animal materials, especially fallen fruits, seeds, roots and tubers. May enter vegetable gardens where can cause extensive damage. Adult male generally solitary except when mating, but females and young may form herds of 20 or more. In Malaysia, females breed throughout the year, producing a litter of up to eight young as frequently as eight months apart. Female is able to breed at only eight months of age, but male does not breed until much older, when strong enough to compete for mates. Female builds a nest of leaves, grass and branches to shelter the young. Pig tracks have two large hoof marks, generally more rounded than those of deer, and usually with the two smaller hind toes leaving marks as well. In South-east Asia, Wild Pig occurs throughout the mainland, and on Sumatra and Java but not Borneo.

Gerald Cubitt

BEARDED PIG *Sus barbatus* SH 90cm

WWF Malaysia/Rodney Lai

Bearded Pigs vary from pinkish-brown to yellow-grey, which may be affected by the colour of the mud in which they have been wallowing. They have a big 'beard' of bristles on the lower jaw, as well as two pairs of fleshy warts or protuberances with upward pointing bristles above each side of mouth. Diet includes fallen fruits and seeds, roots, herbs and other plant material, as well as earthworms and other small animals. Adults normally weigh 55–80kg, but may weigh up to 120kg when food is plentiful. In extensive forest, may form large herds of several hundred individuals that travel long distances in search of food. Females make a nest of saplings and shrubs before giving birth to a litter of up to 11 young. This species is restricted to peninsular Malaysia, Sumatra, Borneo and the Philippines. Javan Warty Pig *S. verrucosus*, found only on Java, is similar but lacks a beard and has three pairs of warts. Heude's Pig *S. bucculentus*, which was thought to resemble Javan Warty Pig, has been described from the Annamite mountains of Laos and Vietnam, but there is some doubt as to whether it is a legitimate species. Its appearance in life has not been described.

LESSER MOUSEDEER *Tragulus javanicus* SH 20cm

WWF Malaysia/Slim Sreedharan

Lesser Mousedeer is one of the smallest ungulates in the world, weighing only 1–2kg. Upperparts generally reddish-brown, underparts white with dark inverted V on throat. The legs are small and delicate, and it walks with hunched appearance and head down, unlike larger deer. Possess neither horns nor antlers, but has canines in both jaws, with upper ones elongated and used for fighting. Lives in understorey of tall and secondary forest, where it is active by both day and night. Diet includes fallen fruits, young shoots, leaves, buds and fungi. Usually solitary, coming together only to breed. Female apparently drums hind feet on ground as a signal. Breeds year round, giving birth to single young, rarely twins, after gestation of four to six months. Female ready to mate within 2 days of giving birth. Young are full grown and ready to breed at 5 months. Ranges from central and southern Thailand, Laos and Vietnam, to Cambodia, peninsular Myanmar, Malaysia, Sumatra, Java and Borneo. Greater Mousedeer *T. napu* is much larger (SH 30–35cm, 4–6kg when full grown), with more mottled grey-brown back and usually shows extra dark mark on side of throat. Range similar to Lesser Mousedeer, but not as far north on mainland and not on Java. Poorly known Silver-backed Mousedeer, *T. versicolor* is small with a speckled back and occurs only in Vietnam.

SAMBAR *Rusa unicolor* SH 140–160cm

Male in velvet (top); female and young (bottom)

Most widespread and common large deer in the region. Skin dark brown with coarse, short dark brown hair; belly darker, inside of legs paler. The ears are broad and the tail is relatively long and mostly black. Adult males develop antlers 50–100cm in length, usually with three tines (branches), one at base, and two at tip. Occurs in a wide variety of wooded habitats from dense rainforest to open deciduous forest and secondary forest. Most active at dusk and at night, resting during day in thick vegetation. Diet consists mainly of twigs, leaves, vines, buds, fallen fruit and grass. Frequently visit salt licks, especially when males are growing new antlers, which they do annually. During the mating season, males fight for territories, and then mate with any females that enter their territories. Up to eight females may enter a territory. A single fawn is born after an eight month gestation. Ranges from India to China, throughout South-east Asia, including Sumatra, Java, Borneo and the Philippines.

JAVAN RUSA *Rusa timorensis* SH 100–110cm

Alain Compost

Alain Compost

Male (top); female and young (bottom)

Similar to Sambar *R. unicolor*, differing in smaller size; more extensive pale underparts, thinner tail, (same colour as back, with tuft of hair at tip), and different-shaped antlers, with outer branch of top fork normally much smaller than inner, instead of longer as in Sambar. Occupies a wide variety of habitats from forests and plantations to open grasslands. Mainly nocturnal, but may be active during the day, feeding on grasses, herbs and shrubs. This species can swim well. Animals living near coast may drink sea water to obtain salt. Females and young form small herds, while males remain in bachelor groups or alone, except during mating. The fawns are unspotted. Calls include an alarm bark and a shrill roar by males during mating season. Range extends from Java and Sulawesi through many of the smaller Indonesian Islands. Introduced into New Guinea, Australia and other regions. Some authorities suggest this deer was native only to Java and Bali, and was introduced to other islands by man.

ELD'S (BROW-ANTLERED) DEER *Rucervus eldii* SH 115–130cm

Gerald Cubitt

A medium-sized, brownish-red deer with paler underparts, a short brownish-red tail and relatively small ears. The antlers are distinctive in having brow tine (branch) curving forward as an extension of main branch, so that antlers appear to be shaped like a bow. Main branch may have several small tines at tip. Eastern Indian and Burmese populations of this species live mainly in low-lying swamps, while those in Thailand and Cambodia use mainly dry deciduous forests. Graze mainly grass, but do eat some leaves, twigs and fruit. May aggregate into herds of up to 50 individuals, but adult males are solitary except during mating season, when they join the herds and fight for females. In the wild, the mating season is apparently between February and May, with single fawn born about eight months later. Formerly widespread in mainland South-east Asia north of the peninsula, this species is now restricted to isolated herds in eastern India and Myanmar, in dry deciduous forests in parts of Cambodia and possibly adjacent countries, and on Hainan Island, China. The remaining populations are endangered by habitat loss (logging and conversion of swamps to agriculture) and illegal hunting. Schomburgk's Deer *R. schomburgki* which had very elaborate antlers with between five and 15 tines on each side, formerly occurred in swampy grasslands in central plains of Thailand and possibly adjacent areas, but these have been largely converted to rice cultivation, and the species is believed to be extinct.

HOG DEER *Axis porcinus* SH 65–72cm

L Bruce Kekule

Gerald Cubitt

Male (top); female (bottom)

Resembles Sambar *Rusa unicolor* but is considerably smaller with greyish to dark brown fur that varies seasonally in colour. Antlers are similar in shape to those of Sambar, but much shorter and thinner, with short tines. Fawns have rows of white spots that are sometimes retained in adults, although other adults lack spots. Typical habitat was formerly flood plains along rivers and marshes with tall grasses, but much of this habitat in Thailand and Indochina has now been converted to rice cultivation. Formerly formed large groups of several dozen individuals, but those deer that remain are largely solitary. Pairs come together to mate in September and October, and fawns are born about eight months later, at beginning of rainy season. The tendency to run through and under low underbrush, like a hog, rather than jumping and bounding like many other deer gives this species its name. Graze mainly on tall grasses along rivers. Range extends from India through Myanmar, Thailand, Laos, Vietnam and Cambodia, but populations in South-east Asia have been greatly reduced by habitat loss and hunting, and are now gone from almost all areas.

RED MUNTJAC *Muntiacus muntjak* SH 50–57cm

Male in velvet (top); female (bottom)

Muntjacs or barking deer are small deer with short antlers joined on a very long pedicel or bony base. The antlers are usually 10–15cm long, with short tine (branch) no more than 5cm long. Female lacks antlers and instead has small bony knob with tuft of hair where male's pedicel would be. Both sexes have canines, elongated in male and used for fighting for mates or for attacking enemies. Fur yellow-brown to chestnut, with variable pattern and tone, darker above and paler below. Face usually shows a black line on muzzle and on inside of antler pedicels. The tail is red above and white below. Both sexes give a loud barking call, repeated for up to an hour if danger threatens, that may carry for up to one km, and probably serves as a warning. Diet mainly leaves and twigs, but will also eat fallen fruit or seeds. In some areas, active during day, but where heavily disturbed, become largely nocturnal. In Thailand, mating takes place in December–January, the single young, rarely twins, born after six month gestation. The most widespread muntjac species, occurring from India to southern China, and throughout South-east Asia including Sumatra, Java and Bali.

BORNEAN YELLOW MUNTJAC *Muntiacus atherodes*
SH 45–50cm

Fletcher and Baylis

This species is slightly smaller than Red Muntjac *M. muntjak*, with paler fur and smaller antlers. Upperparts are yellowish-red with darker brown along middle of back; underparts pale yellowish or white. Tail dark brown above, white below. Males have tiny, unbranched antlers only 1.5–4cm in length, on relatively slender pedicels. The join is relatively smooth, lacking burr of Red Muntjac. Antlers are apparently retained year-round, unlike in other muntjacs. Both sexes give loud barking call. Active mainly during day, feeding on herbs, young leaves, grasses, fallen fruits and seeds. Found only on Borneo, where occurs throughout island, sometimes together with Red Muntjac in some places.

ANNAMITE MUNTJAC *Muntiacus truongsonensis*

W Robichaud/WCS

This small dark muntjac, known only from hill forest in the Annamite Mountains in Laos and Vietnam, was not described to science until 1998. It has blackish fur, a short tail with white edges, short narrow pedicels, and short, unbranched antlers. Skull medium-sized, but body measurements not available. Roosevelts' Muntjac also known only from the Annamites, is genetically quite distinct, but may look similar—identity of animal in photo uncertain. Putao muntjac with chestnut fur, was first described in 1999 from northern Myanmar. Fea's Muntjac with dark brown fur, speckled yellow is found only in western Thailand and adjacent Myanmar. Black Muntjac in northern Myanmar is similar to Fea's, but with black legs and underparts.

LARGE-ANTLERED MUNTJAC *Muntiacus vuquangensis*
SH 67cm

W Robichaud/WCS

Weighing 34kg or more, this large species of muntjac was not described to science until 1994. Described in its own genus, but recent genetic analyses have shown that it is closely related to other muntjacs. Antlers are up to 28cm long, on moderately long, stout pedicels, with well-developed tine (fork) at base, up to 10cm long. Fur is grizzled grey-brown to dark tan, with black line on forehead and white on belly. Tail relatively short and triangular with white underside. Like other muntjacs, male has protruding tusk-like upper canines. Little is known of its ecology and field identification, as it has been rarely seen. A female killed by hunters when being attacked by a Dhole *Cuon alpinus* in early January, was pregnant with small foetus. Known only from evergreen forest in hills of Annamite mountains in central Laos, Vietnam and Cambodia where at risk from forest loss and hunting.

GAUR *Bos frontalis* SH 165–220cm

Adult male Gaur are magnificent, powerful animals, weighing 650–1000kg, with a high ridge on the back, and muscular shoulders and neck. Both sexes are similar in colour, varying from dark reddish-brown to almost black, with white stockings. The horns grow outwards and curve upwards and slightly forwards at tip. Habitat includes forested hills with grassy clearings. May benefit from limited forest disturbance by humans, provided that some forest is left for shelter, and there is no increase in hunting. Graze along forest edges, in natural or artificial clearings or along river banks, where they eat mainly grasses and herbs, but they appear to browse on shrubs and tree branches more than other cattle. Often visit salt licks to obtain necessary minerals. Form herds of up to 30 individuals, led by a dominant bull. The single young is born after a nine month gestation. Range from India throughout mainland South-east Asia, but numbers greatly reduced by hunting, habitat loss and diseases spread from domestic cattle, and now occur as scattered, remnant populations.

BANTENG *Bos javanicus* SH 155–165cm

Herd

Banteng vary from pale brown in young and females to brownish-grey or black in mature bulls. Legs white below the knees, large conspicuous white patch on rump, and white around muzzle. Feeds mainly along clearings and river banks in forested areas, where grazes on various grasses and herbs, though may also eat leaves and twigs. Formerly active by both day and night, resting during hottest part of day under trees for shade. Now, rarely enter open areas except at night because of excessive hunting pressure. Usually form small herds of under 10 individuals, consisting of adult male with females and young. Excess males may be solitary or in small bachelor herds. Give birth to one or two calves after 10 month gestation, and may become receptive again within two months, allowing them to produce young every year. Ranges through Myanmar, Thailand, Laos, Vietnam, Cambodia, peninsular Malaysia, Java and Borneo (though not Sumatra), but now very rare in most areas. Forest clearance initially led to increased habitat, but conversion of secondary areas to agriculture and excessive hunting have led to declines. Their genetic distinctness is also at risk due to hybridization with domestic cattle. A domestic form of Banteng, known as Bali cattle, is widely kept in many parts of Indonesia. Resembles wild form in colour, but is generally smaller with shorter, thicker horns.

Bull (above); young (below)

WWF Malaysia/Mahedi Andau

WILD WATER BUFFALO *Bubalus bubalis* SH 150–190cm

L Bruce Kekule

Herd (above); female (below)

L Bruce Kekule

Wild Water Buffalo resembles domestic Water Buffalo, but larger, weighing 700–1200kg, quicker and more aggressive, with broader, more spreading horns. Both sexes have horns, and those of males may exceed 120cm in spread, the broadest of any living bovid. Preferred habitat open forests and tall grass areas near swamps and streams where they frequently wallow in mud. Feed mainly on grasses. Mating apparently takes place mainly in October–November, with young being born 10 months later. In good habitat, used to form herds of up to 100 individuals, but remaining herds are now much smaller. Formerly widespread through India and South-east Asia, wild Water Buffalo have suffered from conversion of habitat to agriculture, hunting, and competition and disease from domestic buffalo. Now restricted to a few scattered herds, in India, parts of Thailand, and possibly Cambodia, Laos or Vietnam.

SOUTHERN SEROW *Capricornis sumatraensis* SH 85–94cm

Adult (top); young (bottom)

Serow have relatively short bodies and long legs. Upperparts usually black or dark grey, sometimes with reddish tinge; underparts paler. There is a mane of long shaggy hair along back. Both sexes are similar in size with short, slightly curved horns. There is a distinctive large open gland in front of the eye. Mainly inhabit steep forested areas, including limestone mountains, but also found in lowlands and may be found on offshore limestone islands. Mainly solitary, coming together only to mate. Give birth to single young, sometimes twins, which may stay with the mother for up to a year. Active both by day and night, feeding on a wide variety of plants. Found from Peninsular Thailand to Sumatra. Two other species, Red Serow and Chinese Serow occur farther north on mainland.

LONG-TAILED GORAL *Naemorhedus caudatus* SH 50–70

Gerald Cubitt

Also called Chinese Goral, this species is generally smaller and more slender than Southern Serow *C. sumatraensis*, with thinner neck, and a thin mane. Lacks facial glands of serow. Upperparts dark greyish or yellowish-brown; underparts paler. There is a pale patch on the throat, and the tail is short and bushy. Both sexes have similar short horns. Mainly inhabits forested mountainous areas from 1000m to 4000m in the Himalayas. Can climb very steep terrain with ease. Diet includes mainly grasses, herbs and shrubs along cliff edges. Most active in early morning and late evening, but on cool days can be active throughout day. Often found in groups of between four and 12 individuals, though mature males are often solitary outside mating season. Mating takes place in November or December, with one or two young born about six months later. Young reach sexual maturity in two or three years. Found in the mountains of northern Thailand and northern Myanmar and northeast into China. The closely related Red Goral *N. baileyi* with reddish-brown fur, is found in northern Myanmar and adjacent areas of Tibet and Assam. A third species, *N. goral* occurs farther west in the Himalayas. In many areas, numbers have been reduced by excessive hunting.

SAOLA *Pseudoryx nghetinhensis* SH 80–90cm

W Robichaud/WCS

W Robichaud and Ban Vangbiar/WCS and IUCN

The Saola, the only member of its genus, was first discovered by scientists in 1992 and described in 1993. Its relationships to other bovids are still uncertain. It has long horns, slightly curved backwards, superficially resembling an oryx. Varies from dark brown to rich reddish-brown in colour, without a crest, but with a narrow, dark stripe down back to short tail. Face has a distinctive pattern with whitish stripes above and below eye, on side of face, chin and throat. Legs are black with white bands around ankles, and has a large white patch on lower belly and inside back legs. Little is known of its ecology except that it inhabits dense moist forests in the Annamite mountains in central Laos and Vietnam, where its continued existence is seriously threatened by habitat loss and hunting.

RODENTS Order Rodentia

World-wide, rodents are the most diverse group of mammals with over 1800 species. In the region of this book, more than 170 species are known, with new forms being discovered regularly. They are grouped into Sciuridae with both the tree squirrels (about 40 species in 12 genera) and flying squirrels (25 species in 8 genera), the Muroidea (mice and rats) with more than 90 species in 26 genera and 4 families, Hystricidae (porcupines) with 6 species in 3 genera and the newly discovered Diatomyidae with the single species *Laonastes aenigmamus*. The tree squirrels are diurnal, while others are mainly nocturnal.

GIANT SQUIRRELS *Ratufa* spp. TL 70–85cm

Giant squirrels are among the largest arboreal rodents in the world, weighing up to 1.5kg or more. Four species are known in Asia, of which two are in the region. Black Giant Squirrel *Ratufa bicolor*, with black upperparts including legs and shoulders, pale cheeks and throat, is found throughout the mainland except central Indochina, as well as Sumatra and Java. Sunda Giant Squirrel *Ratufa affinis* varies in colour from creamy-brown all over to dark brown above and pale below, but the thighs and shoulders are always paler than the back. It is found in peninsular Thailand, Malaysia, Sumatra and Borneo. Both species are usually solitary, but may occur in twos or threes. Found mainly in the forest canopy where they feed on fruit, seeds and some leaves. Call is a loud chatter audible for several hundred metres. Build a large spherical nest of twigs in the outer branches of large trees. Give birth to one or two young at any time of year.

L. Bruce Kekule

Alain Compost

R. bicolor (top);
R. affinis (bottom)

113

BEAUTIFUL TREE SQUIRRELS *Callosciurus* spp. TL 30–50cm

AH Shoemaker/Mammal Slide Library

C. prevostii (above & right);
C. notatus (below & bottom)

WWF Malaysia

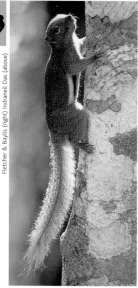

Fletcher & Baylis (right) Indraneil Das (above)

The most familiar and widespread squirrels in the region, with about 14 species occurring. Found in a variety of forested habitats from mature forest to gardens. Prevost's Squirrel *C. prevostii* of peninsular Malaysia, Sumatra, Java and Borneo, is one of the most colourful species, with a striking black, white and red pattern, although some populations in Borneo lack the white, and others are pale grey above. Plantain Squirrel *C. notatus* with a brown back, black and white side stripes and orange belly, has a similar range and is frequently seen in gardens. Several species show striking variation in colour both between different areas and in the same place. The extreme is the Variable Squirrel *C. finlaysoni* which may be pure white, orange or black, or patterned with black above and white below, orange above and grey below and many other combinations. All species are mainly arboreal, feeding on a variety of fruit, seeds, flowers, shoots, leaves, bark and insects. Most species make a large spherical nest of twigs and leaves on branches.

HIMALAYAN STRIPED SQUIRREL *Tamiops mclellandii*
TL 17–21cm

WWF Malaysia/Oon Swee Hock

At least four species of small striped tree squirrels are currently recognized in South-east Asia, with at least one species in most parts of the mainland. Restricted to hill forest in peninsular Malaysia, but occurs widely at lower altitudes farther north. All species have alternating buff and black stripes on the back, although they vary in the relative width of the stripes, overall colour (greyish or orangish) and extent of the buff cheek stripe. Mainly arboreal, in the middle and upper layers of forest trees, but use fruit trees in gardens in northern Thailand. Diet includes many insects as well as fruit and other vegetable matter. Shelter and nest in holes in trees. Solitary or found in small family parties. Calls include a bird-like chirp repeated regularly at one second intervals, and a descending trill of shrill chirps.

SUNDA SQUIRRELS *Sundasciurus* spp. TL 20–23 cm

Five species of *Sundasciurus* occur in the region, most of them brown with buff marks around the eye or on the muzzle. Low's Squirrel *S. lowii*, with the shortest and bushiest tail, is often seen on the ground or in low shrubs or bushes. It is found in peninsular Malaysia, Sumatra and Borneo. Slender Squirrel *S. tenuis*, whose similar range also extends into peninsular Thailand, is more arboreal and has a longer more slender tail. Jentink's Squirrel *S. jentinki* with conspicuous white facial markings, and Brooke's Squirrel *S. brookei* with grey underparts, are confined to the mountains of Borneo. Horse-tailed Squirrel *S. hippurus* has a similar range to Slender Squirrel, and is the largest and most colourful, with reddish-brown upperparts, grey head and tail, and either white or orange belly. Diet of most species includes fruit, seeds and other plant matter.

DG Huckaby/Mammal Slide Library

S. lowii (above);
S. hippurus (below)

Jeremy Holden/FFI

115

BORNEAN MOUNTAIN GROUND SQUIRREL
Dremomys everetti TL 25–32cm

Charles M Francis

This species is distinguished by its relatively pointed muzzle, speckled dark brown upperparts, greyish-white belly and bushy tail that tapers to the tip. It is confined to the mountains of Borneo from 980–3400m. Mainly terrestrial, it may climb trees and feeds on insects, earthworms and fruit. Four other species of *Dremomys* occur in mainland South-east Asia, also mainly in hill and montane forest. The most widespread is Red-cheeked Squirrel *D. rufigenis* with greyish-brown upperparts, greyish-white or buff belly, reddish cheeks and underside of tail and a buff spot behind the ear. It is found in hilly regions of all the mainland countries. Other species vary in the amount of rufous and colour of the underparts, and are restricted to northern Myanmar or Vietnam and adjacent China and India.

GIANT FLYING SQUIRRELS *Petaurista* spp. TL 70–110cm

GS Jones/Mammal Slide Library

P. philippensis grandis

In tall lowland forests, these squirrels can often be seen climbing to the top of their nesting tree, just before dusk, then gliding out across the forest, 100m or more to the next tree on their way to search for food. Like other flying squirrels, the flight membrane stretches between the legs, but the long tail hangs free. Nests are built in hollows in trees, or rarely in crannies in cliffs. Diet consists mainly of fruit with some leaves

P. elegans

and shoots. The Red Giant Flying Squirrel, is one of the largest species (TL up to 110cm), found through much of mainland southeast Asia, as well as Sumatra, Java and Borneo. Distinguished by reddish brown fur with black tips to ears, nose, feet and tail, but some populations are dark. Lesser Giant Flying Squirrel is also found throughout most of the region, but is smaller (TL 70cm), with extensive white spots on a back colour that varies from black (in Borneo) to rufous (mainland) with a black or rufous tail. It is most frequently encountered in the highlands. North of the Isthmus of Kra in Thailand, several additional species occur, including Indian Giant Flying Squirrel which varies from dark grey to brown, frosted with white and a dark tail, Yunnan Giant Flying Squirrel which is reddish except for a grey back frosted white, and Chindwin Giant Flying Squirrel which is plain reddish brown. The precise numbers of species, and the relationships among them are still uncertain, and there may prove to be additional species.

WHISKERED FLYING SQUIRREL *Petinomys genibarbis*
TL 30–36cm

This small flying squirrel is reddish-brown above, speckled grey towards the front, and with a pinkish tinge on the rump. Underparts are cream or orange-buff, and edge of flight membrane is white. Has a diagnostic tuft of long whiskers on the cheek behind each eye. Habits similar to other flying squirrels, sleeping during the day in a nest of leaves and twigs in a tree hole high in the forest, and coming out at night to feed on fruit and other plant material.

WWF Malaysia/HD Rijksen

Found in peninsular Malaysia, Sumatra, Java and Borneo. Several additional species of *Petinomys* occur in the region, with similar shape and flattened tail, but differing in size and coloration.

ARROW-TAILED FLYING SQUIRRELS *Hylopetes* spp.
TL 19–40cm

H. platyurus (top); H. phayrei (bottom)

Seven species of *Hylopetes* squirrels are currently recognized in the region, with generally similar ecology and shape to *Petinomys*, but differences in the skull shape. Three species are poorly known or with very limited ranges: *H. bartelsi* from one location in Java, *H. winstoni* from a single specimen in Sumatra, and *H. sipora* from Sipora Island in the Mentawai Islands. Among the more widespread species, Grey-cheeked Flying Squirrel, *H. platyurus*, found in peninsular Thailand, Malaysia, Sumatra, Java and Borneo, has a brownish back, grey cheeks and a white base to the tail. Red-cheeked Flying Squirrel *H. spadiceus* is similar, but larger, more reddish on the back with reddish cheeks and an orange base to the tail. It occurs in Sumatra, Borneo and mainland South-east Asia into northern Thailand. Pink-cheeked Flying Squirrel, *H. lepidus*, is intermediate in colour and restricted to Java. Phayre's Flying Squirrel *H. phayrei* and the larger Particoloured Flying Squirrel *H. alboniger* are both similar in appearance, with dark brown or grey back fur frosted with white or pale brown, and buff underparts. Both species are found in northern Myanmar, Thailand, Laos and Vietnam, apparently mainly in hill country, though few records are available.

HOUSE MOUSE *Mus musculus* TL 11–19cm

House Mice are associated with towns. Native to Europe and Asia, they have been accidentally introduced through much of the world. The South-east Asian form, often considered a separate species *M. castaneus*, is greyish-brown above with the belly only slightly paler and a long dark tail. It lives mainly indoors, although in areas without other species of mice, it may live outdoors. European forms, which may be darker or lighter (illustrated) sometimes arrive in port cities on ships. Several additional species of mice in the

LL Master/Mammal Slide Library

genus *Mus* occur in the region, mostly living in fields, scrubby areas, or tall grasses in deciduous forests. They are distinguished by a combination of fur colour (many species have white bellies), tail length and colour and skull characters. Some species become pests, eating grain and other human foods, or spreading disease. In Malaysia, mean litter size is four, and the young can reach sexual maturity in 35 days.

SUMATRAN SHREW-MOUSE *Mus crociduroides* TL 20–25cm

Manuel Ruedi

Shrew-Mice are in a separate subgenus from house mice – *Mus (Coelomys)*. They have long pointed muzzles and live in humid forests. The Sumatran species is known only from montane forest in Sumatra. It has brown upperparts, a grey belly with silver-tipped fur, and a tail much longer than the head and body. Javan Shrew-Mouse *M. vulcani* is similar, found in montane forest in Java, but with a shorter tail, and buff-tipped belly fur. Sikkim Mouse *Mus pahari*, in forests of northern Thailand, Laos and Vietnam, has greyer spiny fur. It is nocturnal, and does not dig burrows, but instead makes a nest of dried grass. Little is known of the behaviour or diet in the wild.

RATS *Rattus* spp. TL 20–50cm

R. rattus

Most *Rattus* in the region are associated with humans. Forms of the House Rat *R. rattus* have been accidentally introduced from Europe and Asia to most parts of the world through commercial shipping. Recent studies suggest Asian house rats may be a distinct species, *R. tanezumi*. Their fur is uniformly olive-brown, coarse with few spines, and the tail is all dark. Found mainly near houses, they climb well, and are often found in trees or roof tops. The larger Norway Rat *R. norvegicus*, was probably introduced accidentally from northern Europe or Asia, as it is found only in ports and large cities. The small Polynesian Rat *R. exulans* is another pest species found in houses and cultivated areas. Several additional Rattus species such as Field Rat *R. tiomanicus* and Ricefield Rat *R. argentiventer* live in scrub, secondary forest, crop fields, etc. but rarely enter houses. Other species such as *R. baluensis* on the upper slopes of Mt. Kinabalu in Borneo or *R. osgoodi* from Vietnam live in primary forests.

BURMESE BANDICOOT-RAT *Bandicota savilei* TL 36–40cm

Charles M Francis

Bandicoot-rats have stocky bodies and heads, coarse shaggy fur, and large broad teeth. The Burmese Bandicoot-Rat is found in Laos, eastern Myanmar, central Thailand, Cambodia and Vietnam. It lives in grasses in dry forests, recent clearings and cultivated areas. The Large Bandicoot-Rat *B. indica* (TL 40–70 cm) is found through much of the mainland north of the peninsula, as well as Penang and Java. It forms colonies in rice fields and other recent clearings, burrowing into mounds or termite nests, and feeding on seeds and other plant matter. Its burrowing can severely damage irrigation systems for rice fields. A third species, B. bengalensis, occurs in India and western Myanmar. All three species can be pests, but are also sometimes eaten by humans.

RED SPINY MAXOMYS *Maxomys surifer* TL 35–41cm

Spiny rats in the genus *Maxomys* have relatively larger feet and shorter tails than *Niviventer*. They live in forest or scrub, rarely entering houses. Eleven species occur in the region, differing in size and colour. Red Spiny Maxomys is found throughout the region except northern Myanmar. Its fur, with many stiff spines, is reddish-brown above, orange on the sides and white below. It lives on the forest floor, feeding on plant material such as fruit and roots, as well as insects, slugs and other animals. It lives in underground burrows during the day. Female gives birth to from two to five young.

LH Emmons

NIVIVENTERS *Niviventer* spp. TL 25–40CM

LH Emmons

N. cremoriventer (above); N. fulvescens (below)

About 11 species of *Niviventer* are currently recognized from the region, although the taxonomy is uncertain. Most species of these medium-sized, long-tailed rats are mainly arboreal, although they may forage on the ground. Inhabit lowland or hill forests and rarely enter human dwellings or clearings. One species, *N. hinpoon*, is found only on forested limestone cliffs in Thailand where it may be endangered. Dark-tailed Niviventer *N. cremoriventer* is found in peninsular Myanmar, Thailand and Malaysia, Sumatra, Java and Borneo. Indomalayan Niviventer *N. fulvescens*, is found throughout the mainland north of the peninsula.

Charles M Francis

INDOMALAYAN BAMBOO RAT *Rhizomys sumatrensis*
TL 38–68

L Bruce Kekule

This large stocky rodent (weight 2–4kg) has shaggy, coarse fur and a relatively short tail. The fur is pale brown with dark reddish patches on the crown and cheeks. Bamboo rats have powerful jaws and large feet with sturdy claws, both used for digging extensive underground burrows in stands of bamboo. Live mainly underground during day, but roam above ground at night. Feed on bamboo roots, but may also eat tapioca or sugar cane roots. Range includes much of mainland South-east Asia, as well as Sumatra. Hoary Bamboo Rat *R. pruinosus*, found throughout the mainland to southern China, averages slightly smaller (weight 1–3kg), and has smoother greyish-brown fur with white tips, giving a frosted effect. A third species, with a proportionately shorter tail (less than 24% of head and body, instead of 35–50%) is found from southern China to northern Myanmar and Vietnam. All species are captured for food in some parts of South-east Asia, because of their large size.

LESSER BAMBOO RAT *Cannomys badius* TL 20–34cm

Considerably smaller than *Rhizomys*, weighing 0.5–0.8kg, this species has soft, dense, dark brown or greyish-brown fur, sometimes with a rufous tinge. Often has a white patch on the forehead. The legs are

relatively short; the body is long and sausage-shaped; the tail is short, about 25–35% of the head and body length. These rats use both their teeth and legs to dig extensive burrows under stands of bamboo, in the sides of banks, or under trees, sometimes in very hard ground. The burrow entrance is closed when occupied. Comes out in the evening to feed on a variety of plant matter including shrubs, young shoots and roots. Litter size in captive animals was one or two young. Occurs in central and northern Thailand and Myanmar and Cambodia, often in hilly or mountainous areas.

CJ Philips/Mammal Slide Library

EAST ASIAN PORCUPINE *Hystrix brachyura* TL 70–80cm

A large porcupine, weighing 8–10kg, with long black and white quills. Hairs and quills on back of neck can be raised into a short crest. Long quills on the back are stiff and cylindrical, mostly white with a narrow dark band. A few quills are longer, more slender and all white. Short tail has specially modified rattle quills that are narrow at the base and broader and hollow at the tip. If threatened, porcupines erect their long quills, making them

WWFMalaysia/MPS

appear twice as large, shake their rattle quills and stamp their feet to warn off potential attackers. If still bothered, they charge backwards trying to drive their long quills into the enemy. Quills cannot be thrown, but detach when stuck in an animal, and can cause injury or even death to predators such as dogs or leopards. Largely terrestrial, they feed on roots, tubers, bark and fallen fruit. Dig extensive underground burrows that may be used for many years. Give birth to a single young, rarely twins. Captives may live up to 27 years. Found in a wide range of forested habitats from lowlands to hills throughout mainland South-east Asia as well as Sumatra and Borneo.

JAVAN PORCUPINE *Hystrix javanica* TL 60–70cm

Alain Compost

This species is similar to East Asian Porcupine *H. brachyura* but slightly smaller with no crest and a shorter white tip to the long quills. It is found only on Java and the Lesser Sunda Islands. Sumatran Porcupine *H. sumatrae* and Thick-spined Porcupine *H. crassispinis* of Borneo are both sometimes placed in a separate genus *Thecurus*. They are more brownish than other *Hystrix*, lack a crest and have flattened, grooved quills that become thicker at the tip. The quills are mostly dark with only a short pale tip. Habits are apparently similar to other porcupines.

BRUSH-TAILED PORCUPINE *Atherurus macrourus* TL 48–65cm

WWF Malaysia/ Oon Swee Hock

This species is smaller than *Hystrix* porcupines, weighing about 2.5kg. Upperparts are brown with many relatively short flattened spines interspersed with fewer slender, round pale quills up to 10cm long. Underparts are whitish. Tail is thick and scaly with short bristles along most of its length and a cluster of long dirty white quills at the tip. Mainly terrestrial and nocturnal, this species rests during the day in burrows, caves or crevices between boulders. Small colonies of six to eight may occupy a burrow. Litter size is one. Diet includes roots, tubers, fallen fruit and tree bark. Like other porcupines, may gnaw on bones for calcium. Found through much of mainland South-east Asia as well as Sumatra.

LONG-TAILED PORCUPINE *Trichys fasciculata* TL 50–70cm

WWF Malaysia/LG Saw

Long-tailed Porcupines superficially resemble large rats, but can be distinguished by many short, flat spines, interspersed with a few longer hollow spines, and a thick scaly tail with a tuft of spiny hairs at the tip. Distinguished from Brush-tailed Porcupine *Atherurus macrourus* by shorter spines, smaller size, and relatively longer tail. Some individuals lack tails, suggesting the tail breaks off easily. The smallest of the porcupines (1.7–2.2kg), this species is an agile climber and may feed on leaves and shoots in the tops of trees and shrubs, although it can also be caught in baited traps on the ground. Captive individuals may live more than 10 years. Found in peninsular Malaysia, Sumatra and Borneo.

HARES AND RABBITS Order Lagomorpha

Hares and rabbits are native to most parts of the world except Australasia and Madagascar. In South-east Asia, hares are represented by the Siamese Hare *Lepus peguensis* which occurs in Myanmar, Thailand (except the peninsula), Laos, Vietnam and Cambodia. It is mottled reddish-brown with a pale belly and a black and white tail, and found mainly in open areas. The striped rabbits include one species in Sumatra and another recently discovered in Indochina, and occur in dense forest. In addition, two species of pikas (*Ochotona* spp.) occur in north Myanmar.

STRIPED RABBITS *Nesolagus* spp. TL 36–40

Jeremy Holden/FFI

Sumatran Striped Rabbit (above); Annamite Striped Rabbit (below)

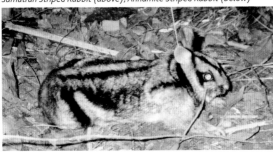

Trinh Viet Cuong/FFI

These small rabbits have a very distinctive fur pattern: buffy-grey with dark brown stripes. The rump and tail are reddish, and the limbs grey-brown. The ears are short, and the tail is small and inconspicuous. They live in the understorey of hill and montane forest, apparently feeding on the succulent stalks and leaves of understorey plants, although captive animals have been known to survive on cooked rice, maize, bread and fruit. Active at night, they rest in burrows during the day, possibly using burrows dug by other animals. Sumatran species is known only from 600–1400m in mountains of Sumatra. The recently discovered Annamite species, from the mountains of central Laos and Vietnam, looks very similar, but genetic analyses suggest it diverged several million years ago. Both species are rare, and potentially endangered by loss of habitat.

FURTHER READING

Cochrane, J. 2000. *The National Parks and other Wild Places of Indonesia*. New Holland Publishers, London.

Corbet, G. B. & Hill, J. E. 1992. *The mammals of the Indomalayan Region: a systematic review*. Natural History Museum/Oxford University Press. A technical work that lists all species in the region with taxonomic details, brief identification tables, and distribution information.

Francis, C. M. 2008. *A Field Guide to the Mammals of South-East Asia*. New Holland Publishers, London. The only comprehensive field guide to the region's mammals, with 72 colour plates, 200 black-and-white drawings, hundreds of maps and detailed descriptions covering all species (nearly 500).

Lekagul, B. & McNeely, J. A. 1977. *Mammals of Thailand*. Association for the Conservation of Wildlife, Bangkok, Thailand. Detailed descriptions of all mammals known to occur in Thailand, mainly with black and white photographs or line drawings.

Medway, L. 1983. *The Wild Mammals of Malaya (Peninsular Malaysia) and Singapore*. Oxford University Press, Kuala Lumpur. Descriptions of all mammal species in peninsular Malaysia, with identification keys and colour paintings of many species.

Nowak, R. M. 1999. *Walker's Mammals of the World, 6th edition*. Johns Hopkins Press, Baltimore, Maryland. 2 volumes. Detailed information on identification and behaviour of all mammals in the world to the genus level, with numerous black and white photographs.

Payne, J., Francis, C. M. & Phillipps, K. 1985. *A field guide to the Mammals of Borneo*. The Sabah Society and WWF, Kuala Lumpur. Colour plates and descriptions of all mammals known from Borneo.

WWF Malaysia. 1999. *The National Parks and other Wild Places of Malaysia*. New Holland Publishers, London.

INDEX